Lecture Notes in Physics

Edited by J. Ehlers, München, K. Hepp, Zürich,
H. A. Weidenmüller, Heidelberg, and J. Zittartz, Köln
Managing Editor: W. Beiglböck, Heidelberg

44

Reinhard A. Breuer

Gravitational Perturbation Theory and Synchrotron Radiation

Springer-Verlag
Berlin Heidelberg GmbH 1975

Author
Reinhard A. Breuer
Max-Planck-Institut für
Physik und Astrophysik
Institut für Astrophysik
Föhringer Ring 6
8 München 23/BRD

Library of Congress Cataloging in Publication Data
Breuer, Reinhard A 1946-
Gravitational perturbation theory and synchrotron radiation.

(Lecture notes in physics ; 44)
Bibliography: p.
Includes index.
1. Gravitation. 2. Perturbation (Quantum dynamics)
3. Synchrotron radiation. I. Title. II. Series.
QC178.B67 521'.1 75-35794

ISBN 978-3-540-07530-1 ISBN 978-3-540-38018-4 (eBook)
DOI 10.1007/978-3-540-38018-4

Originally published by Springer-Verlag Berlin Heidelberg New York in 1975

TABLE OF CONTENTS

LIST OF TABLES AND FIGURES

I. INTRODUCTION

1.1 Gravity Wave Experiments

The existence and some properties of gravitational waves according to (at least the linear approximation to) General Relativity (GR) was suggested by Albert Einstein already two years before the presentation of the final field equations (1913). It is easy to evaluate the weak field approximation of the vacuum equations and to obtain a wave equation for the linearized gravitational wave propagating in Minkowski spacetime with the speed of light. All alternative relativistic theories of gravity - mostly devised to challenge GR - also predict the existence of gravitational waves. There are, incidentally, at present about 50 such theories, but only a few of them are compatible with the present experimental data (for a review see Misner, Thorne & Wheeler, MTW, 1973, Chap. 39). (It appears that only one theory did not predict waves, namely a "stratified" (noncovariant), nonviable scalar theory by Papapetrou (1954), in which the weak field approximation leads to an elliptic differential equation.)

Despite this early proposal of the existence of gravitational waves it has taken more than 50 years to reach some deeper understanding of the nature of gravitational radiation, i.e. to find exact wave solutions of Einstein's equations and to relate the radiation to material sources. Other questions concerned the number of degrees of freedom of the gravitational field, the positive definiteness of its total energy and related to that the nonlocalizability of the gravitational field energy. (A review of those developments and of the experimental situation in GR was given by Brill (1973).)

Because all viable theories predict gravitational waves, the experimental verification of their existence alone will not serve as a test of the validity of GR. This could, however, be achieved by the measurement of the waves' polarization. This question was discussed by Eardley et al. (1973) and will be taken up in Chapter VII of this work. It should be emphasized here that much more needs to be understood about gravitational radiation and its polarization. No exact solution of Einstein's equations which relates radiation to some source is known yet, in other words, no exact model for gravitational radiation has yet been given. Also, no treatment exists for the interaction of the two modes of polarization which is implied by the nonlinearity of Einstein's equations (except in the special case of plane waves where it happens to vanish).

The coupling of gravitational radiation to matter is extremely weak compared to the electromagnetic coupling: the ratio of gravitational to electromagnetic fine structure constants is $\alpha_G/\alpha_e = (G\mu^2_{proton}/\hbar c)/(e^2/\hbar c) \sim 10^{-36}$. For this reason it was thought for a long time that this would preclude the detection of gravitational waves. More recently Weber (1961, 1969, 1970, 1972) proposed and put into operation a detection mechanism whereby part of the wave's energy is tranferred to the mechanical fundamental mode of an aluminium cylinder, by resonance coupling of that mode to the appropriate Fourier-component of the curvature oscillation. The cylinder was chosen to have a mass of 10^6 gr (the maximal size of such cylinders factories could build at the time), a fundamental mode of $\nu = 1660$ Hz, and it has an integrated absorption cross-section ("resonance integral") of $\int \sigma(\nu)\,d\nu = 10^{-21}$ cm^{2}Hz. The measure of the ringing time, the so called "quality factor" Q is $Q = \omega\tau/2 \sim 10^5$, where τ is the decay-time of the barvibration due to an excitation. Piezo-crystals are attached to the bar near the centre where they can tap the maximal bar vibration in the lowest mode. The bar is sensitive to direction, and also to polarization.

Weber (1969, 1972, 1973) has been reporting coincident excitations of his two detectors continuously since 1969. The groups working with bars of Weber's type usually record on magnetic tape the components x and y of the state vector of the observed mode, whereas Weber usually recorded only the squared derivative $(\Delta E)^2$ of the energy $E = x^2 + y^2$. One looks for threshold crossings of $(\Delta E)^2$ or other signal functions computed from x and y. Then one counts coincidences between two or more distant detectors at real time and compares with the number of coincidences at artificially introduced time delays. An excess at zero time delay would be evidence for an external source exciting the detectors simultaneously. Weber still reports results in favour of such evidence, but so far no other group has confirmed it, although the sensitivity of most other detectors is higher than that of Weber's in 1969 - 70. The detectors at Bell Laboratories & Rochester University and those at Munich & Frascati (the latter has recently been moved to Garching near Munich) are even more sensitive than those used by Weber in 1973 - 74. Other detectors of Weber's type have been built at Paris, Moscow, Glasgow and at IBM. In Fig. 1 a comparison of the present relative sensitivities of gravity telescopes is displayed as estimated by Tyson (1973). They are plotted as a function of the frequency range at which the various cylinders operate.

At the seventh international conference on relativity and gravitation (GR7) in Tel Aviv (Summer 1974) the Munich-Frascati group reported a null result from 150 days of common evaluation. Assuming a given pulse strength E one can find the rate of arrivals which would have caused a peak of three standard deviations at zero time delay (in the histogram of coincidence-number versus time-delay, for thresholds chosen optimally for the detection of each pulse strength). The upper limits, for the average daily rate of pulses of favorable direction and polarization, found in this way during 150 days were as follows (Billing et al. 1975):

Pulse Strength E	Daily Rate
0.5 kT = 50 GPU	no pulse in 150 days
0.25 kT = 25 GPU	0.1 per day
0.1 kT = 20 GPU	10 per day
0.025 kT = 5 GPU	500 per day
0.01 kT = 2 GPU	3000 per day

To connect the unit kT with a gravitational pulse strength one has to know that 1 kT in the Munich cylinder (or in Weber's) corresponds to a spectral density of about 2×10^7 erg/cm^2 Hz (at 1660 Hz) in the gravitational wave. (Misner, 1974, defined 1 Gravitation-wave Pulse Unit $\equiv$ 1 GPU = 10^5 erg/cm^2 Hz.) The sensitivity achieved at the Munich detector is such that with a few detectors of this kind a pulse of 1/40 kT would be marginally detectable if there were independent evidence for the time of arrival, e.g. through other astronomical observations. Such a pulse would correspond to the isotropic radiation of 2 % of a solar mass in a bandwidth of 1 kHz at a distance of 1 kpc.

The discrepancy between Weber's results and those of the other groups has not yet been explained satisfactorily, and also no possible source for Weber's pulses has been suggested so far.

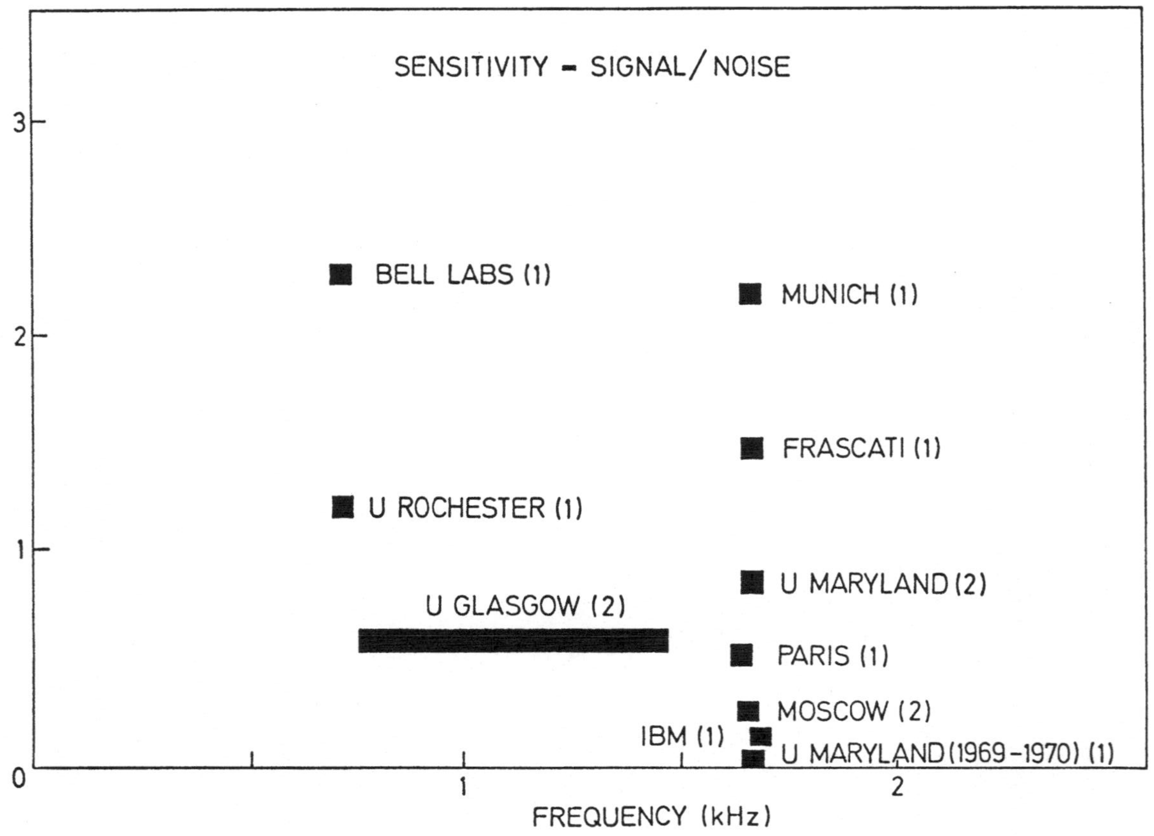

Figure 1. Relative sensitivities to bursts of gravitational radiation of various antenae now operating, assuming equal amplifier noise and similar data analysis. The Rochester & Bell Laboratory gravity telescopes have a fundamental mode at 760 Hz; Maryland and others work at 1660 Hz. The sensitivities drawn are only approximate in the sense that different authors use slightly different definitions of sensitivity. (Figure adapted from Tyson 1973)

Other groups, though not yet in operation, are trying to improve the sensitivity by using temperatures in the range of 10^{-3} K, super-cooling, Laser interferometry or combinations of those (e.g. Stanford). Braginsky from the Moscow group has also suggested the use of dielectric monocrystals, free from dislocations, instead of the usual (polycrystal) aluminum.

Sensitivity improvements of $\sim 10^7$ times the present one are needed to make feasible the observation of gravitational waves from binary systems or pulsars in our galaxy, supernovae, explosions in distant quasars or gravitational collapses up to distance including the Virgo Cluster of galaxies ($\sim$ 10 Mpc away). (Such distances are required to provide a reasonable frequency of events.) Such improvements would indeed open up the field of gravitational wave astronomy (see Press & Thorne, 1972, and Misner, 1974, who gives a list of observational targets & goals for gravitational wave astronomy); milligrad-temperature-technique and Laser-interferometry are expected to be achieved by $\sim$ 1980.

1.2 Motivation of The Gravitational Synchroton Hypothesis

The search for efficient mechanisms for the emission of gravitational radiation was motivated by estimates on energy losses inferred from signals that Weber claimed to have observed since 1969. According to Weber's assumptions the source, situated presumably at the galactic center, should radiate isotropically in a broad band of frequencies. This, combined with an experimental detection efficiency of $\sim$ 20 %, gives an energy loss of roughly $(10^3-10^6)\ M_\odot c^2$ per year (Kafka & Meyer, 1972). As the mass of the galaxy is $\sim (10^{10}-10^{11})\ M_\odot$ a constant loss rate of the above order would empty the galaxy within 10^7-10^8 years. Since the galaxy is about 10^{10} years old this is impossible. Or the source has only "recently" started to radiate, which would imply that we live in a preferred epoch of galactic evolution. In both cases our present period would be distinguished in an inplausible Ptolemean way. In addition, such an energy loss rate should reflect in changes in the motion of stars and hydrogen gas of the galaxy. From observations of stellar motions near the sun and the radial expansion of interstellar hydrogen (studying the 21 cm line) Sciama, Field & Rees (1969) claim that the mass loss rate for the entire galaxy should not exceed $200\ M_\odot c^2/\text{yr}$. Also, no corresponding optical or radio events were detected at the galactic center.

Because of these observational restrictions the following possibilities arise:

(a) The sensitivity of Weber's detectors was under-estimated,

(b) the radiation is not isotropic,

(c) the energy is radiated in narrow band (broad bandwidth assumption wrong), centered around the observed frequency interval.

Possibility (a) was shown to be untrue (Tyson 1973, Kafka & Meyer 1975); in fact, the sensitivity of Weber's detectors was over-estimated which made matters worse. If the radiation were to occur on a narrow frequency band, then no source model known so far could produce this effect. Other drastic mechanisms like collisions within relativistic clusters of black holes (Kafka 1970; Bertotti & Cavalieri 1971) are unable to reduce the energy loss rate. Therefore focus on possibility (b).

There it is of importance to note that the Sun is only (4 ± 12) pc off the plane of the galaxy (Gunn, Kerr & Westerhout 1960) while at a distance of ~ 8 kpc from its center. Any model incorporating anisotropic radiation should produce emission of radiation preferentially along the rotation axis or in the galactic plane. Since only in the latter case direct observations could be made on Earth, the first possibility will not be discussed here. A source at the galactic center which beams radiation within $\Delta\vartheta \sim 10^{-3}$ of the galactic plane would still be observable at the Earth. The high degree of anisotropy in emission processes of such a source would reduce the required energy loss therefore to $\sim (1 - 10^3) M_\odot c^2/\mathrm{yr}$. Hence an astrophysical source is sought which strongly beams radiation into the galactic plane.

From Electromagnetism we know that relativistic charged particles emit synchrotron radiation which can be characterized by the following properties (see, e.g., Sokolov & Ternov 1968):

(a) electromagnetic synchrotron radiation is beamed into an angle of half-width $\Delta\vartheta \sim \gamma^{-1}$, where $\gamma = E/(cm^2)$ is the energy per unit rest mass of the particle; $\vartheta = \theta - \pi/2$ is the latitudinal angle measured off the plane of the orbit at $\theta = \pi/2$.

(b) When the particle is in circular orbital motion with frequency ω_0, then the frequency ω of the radiation generated is a high harmonic of

the orbital frequency: $\omega = m\,\omega_0$, where m is the magnetic multipole number. Above a certain critical frequency $\omega_{crit} = m_{crit}\,\omega_0$ the power spectrum is exponentially damped, where $m_{crit} \sim \gamma^3$.
(c) Electromagnetic synchrotron radiation has a high degree of linear polarization parallel to the plane of the orbit.

Therefore Misner (1972a,b) suggested that an analogous gravitational synchrotron effect might be responsible for Weber's data. A corresponding source emitting gravitational synchrotron radiation (GSR) beamed into the galactic plane would consist of some small body circling relativistically a large compact object at the galactic center. The acceleration would be optimal in the gravitational field of a rotating black hole. Lynden-Bell (1969) has argued that old collapsed quasars typically would be expected to lie at the center of galaxies and hence a single massive black hole might exist at the galactic center with a mass $M = (10^7 - 10^9)\ M_\odot$. A body of such a size would have its angular momentum vector aligned with the galactic rotation and on average accreting matter would move in it's equatorial plane. Hence there would be a natural preference for radiation into the galactic plane from bodies made relativistic by the black hole's field. A body radiating in the fundamental (quadrupole) mode while orbiting a larger body of mass M would radiate at a frequency $\omega \sim \omega_0 \sim M^{-1}$ (in geometrical units) so that maximally a $20M_\odot$ black hole could be allowed to match the frequency at which the experiments operate, $\omega_{exp} = 10^3$Hz. One would, however, not expect such a small black hole (small on a galactic scale) to support strong radiation processes over a time comparable with 10^{10} years.

If the body orbits a massive black hole instead, then

$$\omega_0 \sim M^{-1} \sim 10^{-4}\ \text{Hz}\ (10^8 M_\odot/M)\ ,$$

so that the radiated frequency has to be a high harmonic of the fundamental frequency to be observable at $\sim 10^3$ Hz. In the electromagnetic case a typical harmonic number satisfies

$$\omega_{crit} = m_{crit}\,\omega_0 = \gamma^3\omega_0 = 10^{-4}(10^8 M_\odot/M)\,\gamma^3 \overset{!}{\simeq} 10^3\ \text{Hz}$$

which yields

$$\gamma \simeq 10^2 \left(\frac{M}{10^7 M_\odot} \right)^{1/3}$$

as the energy required for the body. The corresponding beaming angle would be

$$\Delta\vartheta \sim \gamma^{-1} \simeq 10^{-2} \left(\frac{10^7 M_\odot}{M} \right)^{1/3} .$$

A model for GSR with the above properties is discussed in Chap. III; it does not give, however, an astrophysical possible source for GSR, mainly due to the fact that even rapidly rotating black holes are unable to accelerate particles to the relativistic regime. Only if the body is artificially endowed with a relativistic speed and sent around a black hole on certain unstable orbits, then GSR will take place.

1.3 Summary

This article presents methods and results for a gravitational perturbation theory which treats massless fields as linearized perturbations of an arbitrary gravitational vacuum background spacetime. The formalism is outlined for perturbations of type $\{22\}$ spacetimes. As an application, high-frequency radiation emitted by particles move approximately on relativistic circular geodesic orbits is computed. More precisely, the test particle assumption is made; throughout it is therefore assumed that the reaction of the radiation on the particle's motion is negligible (Zel'dovich & Novikov, 1971, p. 51). In particular, these orbits are studied in the gravitational field of a spherically symmetric (Schwarzschild-) black hole as well as of a rotating (Kerr-) black hole. In this model (Misner 1972), the outgoing radiation is highly focussed and of much higher frequency than the orbital frequency, i.e. one is dealing with "gravitational synchrotron radiation".

Stimulated by Weber's experiments to verify the existence of gravitational waves, several other focussing mechanisms have been suggested beside the present one (Chap. II). However, none of these account for the anisotropic emission of radiation. The synchrotron model provides such a mechanism. In Chap. III the properties of relativistic geodesic particle motion relevant to the synchrotron model are studied. Scalar (s=0), electromagnetic (s=1) and gravitational (s=2) synchrotron radiation from particles on relativistic accelerated circular orbits in flat spacetime are calculated in Chap. IV. The result for the power spectrum (i.e. power per unit frequency) as a function of the spin s may be summarized by the formula

$$\frac{dP_{out}^{(s)}(\omega)}{d\omega} \sim \left(\frac{\omega}{\omega_{crit}}\right)^{1-2s/3} \exp(-2\omega/\omega_{crit}), \tag{1.1}$$

where ω_{crit} is a suitable cut-off frequency.

Subsequently (Chap. V), the mathematical foundation is laid for the gravitational perturbation theory. A definition of perturbations of spacetimes is given and the various formalisms of gravitational perturbation theory are reviewed, e.g. the Regge-Wheeler formalism, the Regge-calculus, the Newman-Penrose (NP) formalism and its improvement by Geroch, Held & Penrose (GHP). Within the GHP-formalism, the inhomogeneous perturbation equations are decoupled and separated, rederiving results first obtained by Teukolsky (1972). The properties of the angular wave functions ("spin-weighted spheroidal harmonics") are studied, and, in particular, their low-and-high frequency approximations are found.

An important property of waves with spin $s > 0$ is their polarization. In Chap. VII a formalism is developed to describe partially polarized waves which uses the Stokes parameters of electromagnetism suitably redefined for electromagnetic & gravitational radiation travelling in curved spacetime (Anile & Breuer 1974). Using these Stokes parameters, the equations for electromagnetic and gravitational radiative transfer of polarized radiation are derived.

In Chap. VIII, the power spectrum radiated to future null infinity is obtained for radiation emitted by relativistic test particles in the limit of high frequencies ω, i.e. for the case $M\omega \gg 1$, where M is the mass of the black hole. In the high-frequency limit, the homogeneous wave equations for different spin coincide approximately. However, because there is a different coupling of the field to the source for different spin, the spectrum exhibits a behaviour which can be summarized by

$$\frac{dP^{(s)}_{out}}{d\omega} \sim \left(\frac{\omega}{\omega_{crit}}\right)^{1-s} \exp(-2\omega/\omega_{crit}) \quad , \tag{1.2}$$

where s=0,1,2. Simultaneously, there is a spin-independent focussing of the waves into the equatorial plane with semi-angle

$$\Delta\vartheta = |\omega/\omega_0|^{-1/2} \quad , \tag{1.3}$$

where the equatorial plane is at $\vartheta = \theta - \pi/2 = 0$ and ω_0 is the frequency of the particle's orbit. If a , the angular momentum per unit mass of the black hole, is nonzero, the properties of spectra (1.2) and the focussing (1.3) for the Schwarzschild black hole are changed only quantitatively.

The hole's rotation, in fact, serves as an amplification, so that the spectra corresponding to (1.2) can be expressed as

$$\frac{dP^{(s)}_{out}(\omega,a)}{d\omega} = A(a)\,\frac{dP^{(s)}_{out}(\omega)}{d\omega} \quad , \tag{1.4}$$

where $A(a)$ is a spin-independent amplification factor. The half-width of the beam widens slightly as a increases, according to

$$\Delta\vartheta = |\omega/\omega_0|^{-1/2}\,[1 - a^2\omega_0^2]^{-1/4} , \tag{1.5}$$

i.e. the black hole's rotation defocusses the beam. The factor $A(a)$, however, vanishes in the limit $a \to M$, so that for an extreme Kerr black hole a distant observer will receive no synchrotron radiation. The reason for this behaviour as well as the spin-dependence of Eq. (1.2) are discussed in Chap. IX.

Finally, a number of open problems and questions that emerged throughout the article are collected in order to stimulate the reader's fantasy and inspire him towards further work in gravitational perturbation analysis.

II. FOCUSSING MECHANISMS FOR GRAVITATIONAL RADIATION

2.1 Introduction

The sun lies remarkably close to the galactic plane. Gunn, Kerr & Westerhout (1960) estimated the distance out of the plane to be $z_{\odot} = (4 \pm 8)$ pc. Thus as seen from the galactic center the Sun subtends an angle of $\lesssim 10^{-3}$ rad. It is therefore tempting to explain the apparently excessive energy requirements of Weber's observations by postulating that the radiation is extremely anisotropic, and is only observable because of the privileged position of the Earth.

The first model to explain Weber's results using this idea was proposed by Lawrence (1971). In his, and some other models which will be described in this chapter, the assumption of isotropy is dropped, although in most models the source is believed to be at the galactic center. Instead, radiation is assumed to be radiated anisotropically. A substantial part of the observed intensity is then attributed to amplification by focussing due to some compact object, and/or a disk, at the galactic centre (the gravitational lens effect).

The gravitational lens phenomenon was first described in 1936 by A. Einstein (1936). In a note in response to a suggestion by R.W. Mandl, a Czech electrical engineer, Einstein reported that the lenslike action of one star could lead to an intensification of the image of another star suitably aligned behind it. Zwicky (1957), Liebes (1964) and Bourassa & Kantowski (1975) applied this idea to starlight focussed by the galaxy and by unobservable stars as deflectors. The gravitational lens effect due to black holes was explicitly studied by Cunningham & Bardeen (1972), who considered the problem of an optically radiating star in orbit around a Kerr black hole. They found, among other effects, multiple image production. Gunn & Press (1973) have recently calculated double image or ring-like pattern formation induced on starlight by black holes, because of the focussing effect. Winterberg & Phillips (1973) have shown that there is also a gravitational self-lens effect; the optical image of a luminous mass (of the size close to its Schwarzschild radius) is enlarged by a factor $\simeq 2.59$ and there are ringlike intensity enhancements near the limb of the optical image.

This class of focussing mechanisms relies on the existence of a massive object at the galactic centre when employed for an explanation for Weber's observations. The existence of such an object (black hole) in the galactic nucleus has been proposed by Lynden-Bell (1969) and Rees & Lynden-Bell (1971). Additional support in favour of this assumption for this argument was given by Ryan (1972). A black hole of mass at least $10^6 M_\odot$ - $10^7 M_\odot$ would be able to collect our galaxy around it within the lifetime of the universe. This black hole would presumably spin with its axis aligned with that of the galaxy to a high degree of accuracy, because its angular momentum might be produced mostly by accretion of rotating galactic matter. Bardeen (1970) has argued therefore, that a central black hole would have nearly the maximum angular momentum consistent with collapse. The estimates for the mass possible for a black hole range from $10^6 M_\odot$ to $10^9 M_\odot$.

2.2 Lens Focussing

Weak gravitational waves, in the geometrical optics limit, will follow null geodesics of the background geometry. Lawrence (1971), employing this approximation, used the ray-optics formulae found by Liebes (1964), to argue that the gravitational waves claimed by Weber might be of extragalactic origin, but focussed by the galactic centre acting as a gravitational lens. While sufficient intensification was possible, Lawrence concluded that too few sources were correctly located for the effect to be important. Later (Lawrence, 1972), he hypothesized that a massive oblate rotator exists in the galactic centre, the galaxy and the rotator axes aligned. The spacetime employed is that external to an axisymmetric, uniformly rotating mass of perfect fluid in linearized GR, combined with a slow motion approximation (up to first order in v/c), of the matter. Gravitational waves emitted from the interior of the rotator would - under favourable circumstances - be strongly focussed by its gravitational field into the galactic plane. An average intensification at the earth of one order of magnitude was claimed by Lawrence.

In a subsequent paper by Lawrence (1973), written after the synchrotron model had been proposed, the rotator is replaced by a maximally rotating black hole with most of the (corotating) matter concentrated in its equatorial plane in the neighborhood of the horizon at $r = (1+\delta)M$, $\delta \ll 1$. Null geodesic rays leaving the source at small latitudes $\vartheta_o = \theta - \pi/2 \ll 1$ will arrive asymptotically at angles $\lesssim \vartheta_o \sin \ln(\delta^{-1})$, where the angle ϑ_o is measured by an observer in Bardeen's locally nonrotating frame (LNRF) at the emission point. The intensification of radiation at infinity compared to that at $r = M$, as observed by an observer in an LNRF, is of order ten. By the same calculation performed for the Schwarzschild geometry, with the matter in the unstable orbit close to $3M$, Lawrence obtains a slightly larger intensification factor. The rotation of the black hole has a small defocussing effect on the beam width. This result was also found in the wave-equation analysis in the GSR model.

Campbell and Matzner (1973) have also performed the calculation of the focussing of null rays in the Schwarzschild geometry and found the same enhancement effect.

2.3 Disk Focussing

As well as stars and black holes, relativistic disks can also enhance the intensity of radiation in directions near to their planes (Jackson 1973). Jackson suggests "that the galactic nucleus contains a relativistic disk" (flat, strongly self-gravitating systems of material, the collapse of which is prevented by rapid rotation), "whose plane coincides with that of the galaxy, and that gravitational radiation is generated at a 'reasonable rate' within the disk". Jackson then considers an infinitesimally thin Newtonian disk and concludes (for the situation in our galaxy) that rays emerging at the Earth at an angle of $10^{-3} - 10^{-4}$ rad cross the galactic plane 100 - 500 times on the way from the galactic centre. This would provide an enhancement factor of 20 - 40. With general relativistic modifications (relativistic disk, disk rotation) the actual value for the intensification might even be higher.

2.4 Naked Singularities

In a quite different approach, Penrose (1972) has tried to take advantage of the peculiar focussing properties of naked singularities of the Kerr type. Penrose has coined the "Cosmic Censorship Hypothesis", which states that all of the singularities occuring in nature are "clothed", i.e. hidden from the outside world by an absolute event horizon. However, there is still no theorem which excludes a naked singularity as the final product of gravitational collapse. Indeed work by Müller zum Hagen et al. (1973, 1974) has shown that collapse of a dust cloud might lead to a naked singularity.

If a naked singularity of the Kerr type occurs, then as Carter has shown, any effect of it can be observed only in its equatorial plane. Lines of sight "bounce" before they reach the singularity if they are not in this plane. If the naked singularity were emitting gravitational waves, they would be beamed into this plane - which may be the galactic plane - for reasons given above. Penrose, playing advocatus diaboli, has suggested that if this phenomenon somehow applied to the centre of our galaxy, we could imagine the singularity being responsible both for the existence of the galactic plane and for Weber's gravitational waves.

2.5 Discussion

The null-ray methods (geometrical optics approximation) used in the models discussed in this chapter are of course inadequate for genuine wave phenomena and give incorrect results for the lower modes of radiation. For a plane electromagnetic wave being scattered by a sun-like object Herlt & Stephani (1975) have contrasted the different predictions by wave optics and geometrical optics. The latter also ignores backscatter effects caused by the presence of curvature. Hence the above results describe only partially the solutions of the relevant wave equations. In addition, all of these models have to assume the radiation to be emitted anisotropically. The present work, however, will provide an emission mechanism for anisotropic emission of gravitational radiation.

If, however, in addition to a central black hole, a disklike matter distribution with the parameters required by Jackson's model is present in the galactic nucleus, then the disk-lens effect might be able to contribute independently to the beaming of gravitational waves in addition to any other effect that might occur. Taken by itself, neither lens nor disk focussing is able to reduce the mass-loss rate as estimated by Weber to the observational upper limit of (70 - 200) $M_\odot$ per year. On the other hand, most physicists would prefer any reasonable alternative to the naked singularity postulate, for if the cosmic censorship hypothesis breaks down, physicists have no means of predicting the future.

III. RELATIVISTIC GEODESICS

3.1 Introduction

In the models described in the previous chapter it was assumed either that the radiation was emitted anisotropically or that the emission was isotropic at the source, but certain focussing effects would beam a significant amount of the radiation into the galactic plane, the plane where the observation takes place (earth). The synchrotron model has been devised to provide a mechanism for anistropic emission of radiation.

In the familiar case of electromagnetic synchrotron-radiation, charges moving relativistically in flat space emit radiation strongly beamed into the plane of the orbit. If the orbit is circular, then the radiation is emitted at very high harmonics of the fundamental frequency. What will happen to gravitational radiation from a particle in a relativistic geodesic orbit? Strictly speaking, a radiating particle should not move on a geodesic because of radiation damping. There is no contradiction here with the principle of equivalence: the latter is a local statement while radiation reaction is in an essential way a nonlocal phenomenon. An observer in the rest frame of the particle, moving at nearly the speed of light passing a black hole in a small angle scattering orbit, will experience the black hole as a virtual plane gravitational wave (Pirani 1959). This virtual radiation will ne partly deflected by the test particle; a different observer, who "sits still" (in the rest frame of the black hole) at the same point and sees the particle rushing by, will observe the scattered virtual radiation as radiation from the test particle (bremsstrahlung). This method of obtaining radiation by virtual quanta is well known in electromagnetism as the Weizsäcker-Williams method. For the case of gravitational radiation from relativistic particles, the Weizsäcker-Williams method has been generalized by Matzner and Nutku (1974). Although this method has been used only for gravitational bremsstrahlung from small scattering-angle orbits, it should in principle be applicable to the large angle scattering.

Another question arises: when is the particle indeed "relativistic", i.e. when is it energetic enough to be able to radiate strongly? Here we will use the simple criterion that the particle's proper velocity (e.g. as measured by a static observer in the Schwarzschild geometry) has to approach the speed of light. A more useful criteria will turn out to be that the tangent vector of the particle's worldline is asymptotically close to a null direction and this null directio: is not a principal null direction of the background metric (see Chap. IX). In the next two sections it will be seen that only unstable particle orbits can satisfy this condition, even in the case of a maximally rotating black hole. For particles moving on stable circular orbits, the main part of the radiation is in the low frequency modes.

3.2 Geodesics in the Schwarzschild Spacetime

The geodesic equations in the Schwarzschild spacetime are well known and can be found in the literature (see, e.g., MTW 1973). Consider a particle on a geodesic orbit around a black hole of mass M described by the Schwarzschild coordinates the corresponding line element is given by

$$ds^2 = -(1-2M/r)dt^2 + (1-2M/r)^{-1}dr^2 + r^2 d\Omega^2, \qquad d\Omega^2 = d\theta^2 + \sin^2\theta \, d\varphi^2 . \tag{3.1}$$

Without loss of generality the motion can be restricted to the $(\theta = \frac{\pi}{2})$-plane. Then the geodesic equations for metric (3.1) are

$$\left(\frac{dr}{d\tau}\right)^2 + V(r) = \gamma^2 \quad ; \tag{3.2a}$$

$$V(r) = \left(1 - \frac{2M}{r}\right)\left(\frac{\ell_z^2}{r^2} + 1\right) \quad ; \tag{3.2b}$$

$$\frac{d\varphi}{d\tau} = \frac{\ell_z}{r^2} \quad ; \tag{3.3}$$

$$\frac{dt}{d\tau} = \gamma\left(1 - \frac{2M}{r}\right)^{-1} . \tag{3.4}$$

A plot of the potential (3.2b) is given in Fig. 12 since the potential coincides with the high-frequency form of the corresponding potential of the scalar wave equation. Here τ is the proper time and both γ and ℓ_z are conserved quantities;

$$\gamma = E/\mu = -p_t = (1 - \frac{2M}{r}) \frac{dt}{d\tau} \tag{3.5}$$

is the energy per unit rest mass of the test particle as measured by an observer at infinity. The constant

$$\ell_z = \frac{L_z}{\mu} = p_\varphi = r^2 \frac{d\varphi}{d\tau} \tag{3.6}$$

is the conserved angular momentum of the particle. The angular frequency measured by an observer at infinity of the particle motion is given by

$$\omega_o = \frac{d\varphi}{dt} = \frac{d\varphi}{d\tau}\frac{d\tau}{dt} = \frac{\ell_z (1 - \frac{2M}{r})}{r^2 \gamma} \quad . \tag{3.7}$$

Eq. (3.2) can be simplified for circular motion, for then

$$-V(r_o) = 0 = \gamma^2 - (1 - \frac{2M}{r_o})(\frac{\ell_z^2}{r_o^2} + 1) \tag{3.8}$$

$$-\frac{dV}{dr}(r_o) = 0 = \frac{1}{r_o^4}(2\ell_z^2 r_o - 2Mr_o^2 - 6M\ell_z^2) \ , \tag{3.9}$$

where r_o is the radius of the orbit. These last two equations may be solved simultaneously for γ and ℓ_z in terms of r_o . One finds

$$\gamma = (1 - \frac{2M}{r_o})(1 - \frac{3M}{r_o})^{-1/2} \quad , \tag{3.10}$$

$$\bar{b} \equiv \frac{\ell_z}{\gamma} = \pm (Mr_o)^{1/2} (1 - \frac{2M}{r_o})^{-1} \quad . \tag{3.11}$$

The quantity $\bar{b}$ is the angular momentum per unit energy-at-infinity; it is defined for all orbits. A closely related quantity,

$$b = \frac{\ell_z}{(\gamma^2 - 1)^{1/2}} = \frac{\bar{b}\gamma}{(\gamma^2 - 1)^{1/2}} = \pm r_o(\frac{4M}{r_o} - 1)^{-1/2} \ , \tag{3.12}$$

is more convenient for unbound, $\gamma > 1$, orbits, where b is the impact parameter for the scattering orbits into which the unstable circular orbits with $3M < r_o < 4M$ are pushed by small perturbations; for high energies-at-infinity γ, $b \simeq \bar{b}$. No circular orbits exist with $r_o < 3M$; but $r_\gamma = 3M$, where γ becomes infinite, is the radius of an unstable circular photon orbit.

Using (3.10) and (3.11), we may simplify many formulae of interest, i.e., by eliminating γ and ℓ_z and re-expressing the relations as functions of r_o only. In particular, the angular frequency (3.7) becomes ("Kepler's law")

$$\omega_o = \frac{d\varphi}{dt} = \left(\frac{M}{r_o^3}\right)^{1/2} , \tag{3.13}$$

while $u^t = dt/d\tau$ reduces to

$$\frac{dt}{d\tau} = \left(1 - \frac{3M}{r_o}\right)^{-1/2} . \tag{3.14}$$

Although circular particle orbits exist for all radii $r_o > 3M$, not all of these orbits are stable and not all are "bound". Stability occurs if and only if $d^2V(r_o)/dr^2 \geq 0$. From Eq. (3.2b) we see that the last stable orbit is at $r_o = 6M$, where $\gamma = 2\sqrt{2}/3 < 1$ and $\ell_z = 2\sqrt{3}$. All circular orbits with $3M < r_o < 6M$ are unstable. Like in Newtonian theory, the orbits are bound (unbound) when $\gamma < 1$ ($\gamma > 1$), provided the unstable circular orbit at $r_o \simeq r_\gamma$ ($\ell_z \to \infty$) is considered as marginally unbound for $\gamma > 1$.

With the criterion given in the introduction a particle becomes relativistic only when it approaches the photon orbit, which is for $r_o \to r_\gamma = 3M$, where $\ell_z, \gamma \to \infty$. These orbits correspond to unbound high-energy states with impact parameters $b \simeq 3\sqrt{3}\,M$, scattered by the black hole through angles $\Delta\varphi \gg 2\pi$; their frequency $\omega_o \simeq (27M^2)^{-1/2}$. Letting $r_o = (3 + \delta)M$ with $\delta \ll 1$ one finds

$$\gamma^2 \simeq \frac{1}{3\delta} , \quad r_o \simeq r_\gamma \left(1 + \frac{1}{9\gamma^2}\right) , \tag{3.15}$$

and

$$(u^t)^2 = \left(\frac{dt}{d\tau}\right)^2 = 3/\delta . \tag{3.16}$$

3.3 Circular Geodesics in the Kerr Spacetime

Equations of motion in the Kerr geometry have well been studied in the literature (Carter 1968, de Felice 1968, Wilkins 1972; see also discussions in Misner 1969, MTW 1973 , Bardeen et al. 1972, Stewart & Walker 1973). Following those outlines we confine the present analysis to circular, equatorial geodesics and derive those of their properties which are essential for GSR calculations.

In the coordinates of Boyer & Lindquist (1967) the Kerr (1963) metric is given by

$$ds^2 = -\left[\frac{\Delta_K \Sigma^{1/2}}{B^{1/2}} dt\right]^2 + \left[\frac{\Sigma^{1/2}}{\Delta_K^{1/2}} dr\right]^2 + \left[\Sigma^{1/2} d\theta\right]^2 + \left[\frac{B^{1/2}\sin\theta}{\Sigma^{1/2}}(d\varphi - \frac{2aMr}{B} dt)\right]^2 \tag{3.17a}$$

$$\equiv -(\omega^{\hat{t}})^2 + (\omega^{\hat{r}})^2 + (\omega^{\hat{\theta}})^2 + (\omega^{\hat{\varphi}})^2 , \tag{3.17b}$$

where $\Delta_K = r^2 - 2Mr + a^2$, $\Sigma = r^2 + a^2\cos^2\theta$ and $B = (r^2 + a^2)^2 - \Delta_K a^2\sin^2\theta$; θ and φ are standard, spherical polar coordinates at infinity, and a is the black hole's angular momentum per unit mass subject to $0 \leq |a| \leq M$. At the larger root of $\Delta_K = 0$ or $r_h = M + (M^2 - a^2)^{1/2}$ this metric possesses an absolute event horizon; from its interior no signals can return to infin ity. There are two Killing vectors $\partial/\partial t$ and $\partial/\partial\varphi$, representing the symmetries of a stationary, axisymmetric space-time. The vector $\partial/\partial t$ changes from timelike to spacelike at the surface called the static limit (or infinite-redshift-surface),
$r_{ir} = M + (M^2 - a^2\cos^2\theta)^{1/2}$.

Enclosed between the horizon and the stationary limit surface is the so-called ergosphere (Ruffini and Wheeler 1971), in which processes can occur which extract rotational energy of the black hole. The metric (3.17) leads to geodesic equations, e.g. by using the variational principle $\int L d\lambda = 0$, where

$$L \equiv \frac{1}{2}\left[-\frac{\Delta_K \Sigma}{B}\left(\frac{dt}{d\lambda}\right)^2 + \frac{\Sigma}{\Delta_K}\left(\frac{dr}{d\lambda}\right)^2 + \Sigma\left(\frac{d\theta}{d\lambda}\right)^2 + \frac{B\sin^2\theta}{\Sigma}\left(\frac{d\varphi}{d\lambda} - \frac{2aMr}{B}\frac{dt}{d\lambda}\right)^2\right]. \tag{3.18}$$

The Euler-Lagrange equations

$$\frac{d}{d\lambda}\left[\frac{\partial L}{\partial(dx^a/d\lambda)}\right] - \frac{\partial L}{\partial x^a} = 0 \tag{3.19}$$

then yield the geodesic equations. The momenta $p_t = -E \equiv -\mu\gamma$ and

$p_\varphi = L_z = \mu \ell_z$ are conserved quantities because of the existence of the two Killing vectors. They are

$$p_t \equiv \frac{\partial L}{(dt/d\lambda)} = -\frac{\Delta_K \Sigma}{B}\frac{dt}{d\lambda} - \frac{2aMr\sin^2\theta}{\Sigma}\left(\frac{d\varphi}{d\lambda} - \frac{2aMr}{B}\frac{dt}{d\lambda}\right) , \quad (3.20a)$$

$$p_\varphi \equiv \frac{\partial L}{\partial(d\varphi/d\lambda)} = \frac{B\sin^2\theta}{\Sigma}\left(\frac{d\varphi}{d\lambda} - \frac{2aMr}{B}\frac{dt}{d\lambda}\right) . \quad (3.20b)$$

Throughout, μ is the particle's conserved rest mass energy and $\tau = \mu\lambda$ is its proper time, where λ is an affine parameter.

Another constant of motion, related to the existence of a Killing tensor of valence 2 in the Kerr space-time, was shown to exist by Carter (1968). It is, with $\mu = 1$ for particles and $\mu = 0$ for photons

$$K = p_\theta^2 + \cos^2\theta\left[a^2(\mu^2 - E^2) + L_z^2/\sin^2\theta\right] . \quad (3.21)$$

The quantity K appears in the radial and angular geodesic equations. For geodesic orbits remaining in the equatorial plane we have $\theta = \pi/2$, $d\theta/d\lambda = 0$; hence $p_\theta = 0$ and $K = 0$ in the equatorial plane. To find the conditions for circular geodesics we obtain the equations of motion from (3.19) and (3.20) (or, set $\theta = \pi/2$ in Eqs. (33. 32) of MTW):

$$\left(\frac{dr}{d\lambda}\right)^2 + W(r) = 0 , \quad (3.22a)$$

$$r^2\frac{d\varphi}{d\lambda} = \frac{aP}{\Delta_K} - a(E - L_z) , \quad (3.22b)$$

$$r^2\frac{dt}{d\lambda} = \frac{(r^2 + a^2)P}{\Delta_K} - a(aE - L_z) , \quad (3.22c)$$

where

$$W(r) = -r^{-4}\left\{P^2 - \Delta_K\left[\mu^2 r^2 + (L_z - aE)^2\right]\right\} , \quad (3.22d)$$

$$P = \mu\gamma(r^2 + a^2) - aL_z . \quad (3.22e)$$

$W(r)$ is an effective potential governing the radial motion in the Kerr equatorial plane. Circular particle orbits are given by

$$W(r_o) = 0 \quad \text{and} \quad \frac{dW(r_o)}{dr} = 0. \tag{3.23}$$

The solution for $\gamma = E/\mu$ and $\ell_z = L_z/\mu$ (Bardeen, Press and Teukolsky, 1972) is

$$\gamma_\pm = \frac{r_o^2 - 2\,Mr_o \pm a(Mr_o)^{1/2}}{r_o\left[r_o^2 - 3Mr_o \pm 2a(Mr_o)^{1/2}\right]^{1/2}}\,, \tag{3.24}$$

$$\ell_{z\,\pm} = \frac{\pm(Mr_o)^{1/2}\left[r_o^2 \mp 2a(Mr_o)^{1/2} + a^2\right]}{r_o\left[r_o^2 - 3Mr_o \pm 2a(Mr_o)^{1/2}\right]^{1/2}}\,. \tag{3.25}$$

The upper signs refer to prograde orbits (corotating with the black hole); lower signs represent retrograde orbits. For photon orbits, $\ell_z, \gamma \to \infty$ (i.e. the denominators in (3.24) and (3.25) vanish):

$$r_\gamma^2 - 3Mr_\gamma \pm 2a(Mr_\gamma)^{1/2} = 0. \tag{3.26a}$$

The solution is

$$r_{\gamma\,\pm} = 2M\left\{1 + \cos\left[\tfrac{2}{3}\cos^{-1}(\mp \tfrac{a}{M})\right]\right\} \tag{3.26b}$$

for the co- and counterrotating circular, equatorial photon orbit (see Fig. 1). The particle orbits are radially stable if $d^2W(r_o)/dr^2 \geqslant 0$ or

$$6r(\gamma^2 - 1) + 4M \geq 0.$$

For the last stable orbit this expression equals zero:

$$r_{\ell s}^2 - 6Mr_{\ell s} \pm 8a(Mr_{\ell s})^{1/2} - 3a^2 = 0.$$

This equation is satisfied by the relation first found by Bardeen (1970a)

$$3a = \pm\, r_{\ell s}^{1/2}\, M^{1/2}\left[4 - \left(\frac{3r_{\ell s}}{M} - 2\right)^{1/2}\right]\,, \tag{3.27a}$$

or, when inverted for $r_{\ell s}$, by

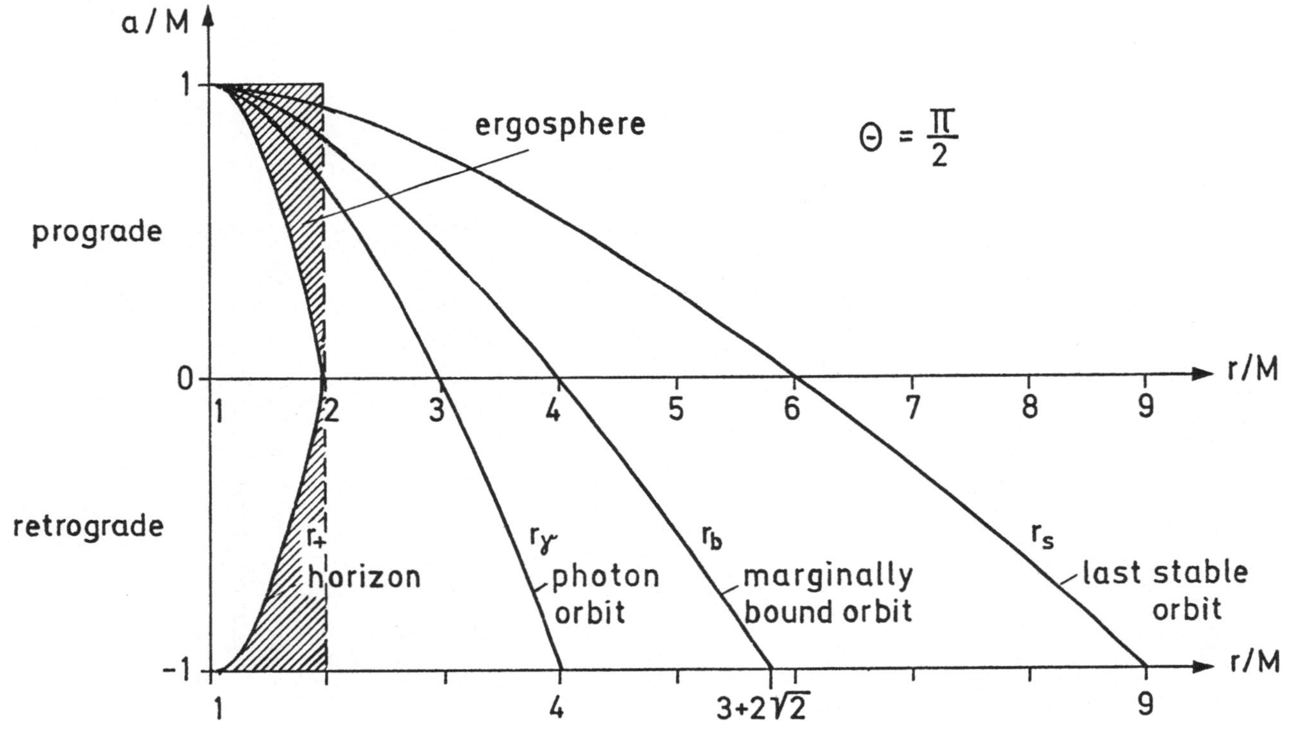

Figure 2. Equatorial geodesics in the Kerr geometry. Negative values of a/M represent retrograde orbits. For $a = M$ all corotating orbit radii seem to approach $a = M$, but in fact the stable orbits do not approach the photon orbit (it is only the coordinate which degenerates). (Graph in parts from Stewart & Walker 1973)

$$r_{\ell s\pm} = M\left\{3 + Z_2 \mp \left[(3 - Z_1)(3 + Z_1 + 2Z_2)\right]^{1/2}\right\}, \tag{3.27b}$$

where

$$Z_1 = 1 + \left(1 - \frac{a^2}{M^2}\right)^{1/3}\left[\left(1 + \frac{a}{M}\right)^{1/3} + \left(1 - \frac{a}{M}\right)^{1/3}\right],$$

$$Z_2 = \left(3\frac{a^2}{M^2} + Z_1^2\right)^{1/2}$$

(see Fig. 1). The last stable orbit is also the circular orbit of minimum energy, because there $\partial\gamma(r_{\ell s})/\partial r = 0$. The orbits are bound for $\gamma < 1$ (Wilkins 1972). The marginally bound orbit occurs for $\gamma = 1$, which gives, with (3.24),

$$r_{b\pm} = \left[M^{1/2} + (M \mp a)^{1/2}\right]^2 . \tag{3.28}$$

For all prograde (retrograde) orbits, $r_{\ell s} > r > r_b > r_h$ holds. For $a = M$ all prograde radii seem to approach M. However, if this limit is evaluated more carefully for an almost extreme Kerr black hole with $a^2 = M^2(1-\alpha^2)$, $\alpha^2 \ll 1$, then

$$\begin{aligned} r_h &\simeq M\left[1 + \alpha\right], \\ r_{\gamma+} &\simeq M\left[1 + 2\,\alpha\,/3^{1/2}\right], \\ r_{b+} &\simeq M\left[1 + 2^{1/2}\alpha\right], \\ r_{\ell s+} &\simeq M\left[1 + (2\alpha^2)^{1/3}\right]. \end{aligned} \tag{3.29}$$

Therefore there is a coordinate degeneracy at $r = M$ in the extreme Kerr spacetime and the various radii do not coincide. In fact, even for $a = M$ there is a nonzero proper distance between the last stable orbit, the circular photon orbit and the horizon, respectively.

For $r_\gamma < r_o < r_{\ell s}$ the circular orbits are unstable. For $r_o < r_\gamma$ no circular geodesic (stable or unstable) exists. The orbital frequency

$$\omega_o = \frac{d\varphi}{dt} = \frac{d\varphi}{d\tau}\frac{d\tau}{dt}$$

can be calculated by inserting the expressions for ℓ_z and γ into Eq. (3.22a,b) to get $d\varphi/d\tau$ and $d\tau/dt$. One obtains

$$\frac{d\varphi}{d\tau} = \frac{M^{1/2}}{r_o^{3/4}\left[r_o^{3/2} - 3Mr_o^{1/2} \pm 2M^{1/2}a\right]^{1/2}} \quad , \tag{30.a}$$

$$\frac{dt}{d\tau} = \frac{r_o^{3/2} \pm M^{1/2}a}{r_o^{3/4}\left[r_o^{3/2} - 3Mr_o^{1/2} \pm 2M^{1/2}a\right]^{1/2}} \quad , \tag{3.30b}$$

and hence

$$\omega_o = \pm \frac{M^{1/2}}{r_o^{3/2} \pm M^{1/2}a} \quad . \tag{3.31}$$

The orbits become relativistic when $r_o \to r_\gamma$ or $\gamma \gg 1$. Let $r_o = r_\gamma(1+\delta)$ with $\delta \ll 1$ as in Eq. (3.15), then one finds from Eqs. (3.22c), (3.30b) for prograde energetic orbits

$$\gamma^2 \simeq \frac{r_\gamma - M}{6r_\gamma M} \quad , \quad r_o \simeq r_\gamma\left(1 + \frac{r_\gamma - M}{6r_\gamma \gamma^2}\right) \quad . \tag{3.32}$$

Energetic orbits require $\delta \ll r_\gamma/M - 1$, a slightly stronger condition than $\delta \ll 1$. For stable orbits, γ remains finite for $a \to M$; in fact from Eq. (3.29) one can show that in this limit

$$\ell_z \to 2M/\sqrt{3} \, , \qquad \gamma \to 1/\sqrt{3}$$

for the last stable orbit $r_{\ell s}$, so that particles executing even extreme stable circular motion are never highly energetic.

A particle infalling from large radii will reach the high-energy unstable orbit for certain values of the impact parameter $\bar{b} = \ell_z/\gamma$, namely, when $\bar{b} = \ell_{z+}/\gamma_+$ or $\bar{b} = \ell_{z-}/\gamma_-$, where $\ell_{z\pm}$, $\gamma_\pm$ are given by equations (3.24) and (3.25); i.e., the high-energy particle almost acts like a photon. In terms of $b = \ell_z/(\gamma^2 - 1)^{1/2}$ a particle is captured by the black hole if it is aimed from infinity with $\gamma \gg 1$ and

$$\left.\frac{\ell_{z-}}{(\gamma_-^2 - 1)^{1/2}}\right|_{r=r_{\gamma-}} \leq b \leq \left.\frac{\ell_{z+}}{(\gamma_+^2 - 1)^{1/2}}\right|_{r=r_{\gamma+}} , \qquad (3.33)$$

When in Eq. (3.33) the equal signs are used, it will assume a pro- or retrograde unstable circular orbit, respectively. Otherwise the particle reaches a perihelion and escapes again.

3.4 Locally nonrotating Observers in the Kerr Spacetime

In the Kerr spacetime it turned out to be extremely useful to distinguish a specific family of observers, i.e. a certain timelike congruence for which spacetime appears particularly "simple". This is the notion of locally nonrotating frames[+] as introduced by Bardeen (1970b).For discussions see also Bardeen et al. (1972), Stewart & Walker (1973) and MTW.

In flat spacetime there exists always a preferred timelike congruence for which the local properties of spacetime do not change. These curves are trajectories of a one-parameter group of isometries and have the properties to be geodesic (inertial) and hypersurface-orthogonal, i.e. no Coriolis-forces are present with respect to them. In the Kerr spacetime no timelike congruence exists which has all those properties simultaneously. If one renounces the condition of geodesy one can at least find a timelike congruence which is hypersurface-orthogonal. For generalizations of this "restframe" to stationary spacetimes see Heunig, Schücking & Vishveshwara (1974). The congruence may be characterized by demanding that its conserved orbital angular momentum $\ell_z = p \cdot \frac{\partial}{\partial\varphi} = 0$. Despite this fact, this congruence has non-zero angular velocity relative to infinity and is not geodesic. Note that angular momentum and angular velocity are not proportional in the Kerr spacetime. These observers have the angular velocity

$$\omega_B \equiv - g_{\varphi t}/g_{\varphi\varphi} = 2Mar/B , \qquad (3.34)$$

[+] Warning: Whereas "frame" frequently refers to a tetrad, here it refers to a congruence of timelike lines.

and basis vectors for them are, at the point (t,r,θ,φ),

$$e_{\hat{t}} = \left[\frac{B}{\Sigma\Delta_K}\right]^{1/2}\left[\partial/\partial t + \omega_B \partial/\partial\varphi\right], \qquad e_{\hat{\varphi}} = \frac{\Sigma^{1/2}}{B^{1/2}\sin\theta}\,\partial/\partial\varphi ,$$

$$e_{\hat{r}} = \left[\frac{\Delta_K}{\Sigma}\right]^{1/2}\partial/\partial r, \qquad e_{\hat{\theta}} = \Sigma^{-1/2}\partial/\partial\theta . \tag{3.35}$$

The dual basis forms $\omega^{\hat{a}}$ can be directly read off from the metric (3.17) to be

$$\omega^{\hat{t}} = \left[\frac{\Sigma\Delta_K}{B}\right]^{1/2} dt , \qquad \omega^{\hat{\varphi}} = \left[\frac{B}{\Sigma}\right]^{1/2}\sin\theta\,\left[d\varphi - \omega_B\,dt\right] ,$$

$$\omega^{\hat{r}} = \left[\frac{\Sigma}{\Delta_K}\right]^{1/2} dr, \qquad \omega^{\hat{\theta}} = \Sigma^{1/2}\,d\theta . \tag{3.36}$$

The physical reason for introducing such a tetrad is that it absorbs, as much as possible, the socalled dragging of inertial frames due to the rotation of the black hole. To make this clearer, let us summarize some of the (not necessarily disjoint) properties of LNRFs.

(a) Observers in the LNRF have as world lines

$$r = \text{const}, \quad \theta = \text{const}, \quad \varphi = \omega_B t + \text{const};$$

(b) particles or photons with zero angular momentum will circle the black hole with frequency ω_B; i.e. trajectories which start radially at infinity will stay radial in a LNRF, as

$$p_{\hat{\varphi}} = e^{a}_{\hat{\varphi}}\, p_a = \frac{\Sigma^{1/2}}{B^{1/2}\sin\theta}\,\ell_z = 0 . \tag{3.37}$$

(c) A LNRF observer has a worldline perpendicular to the $t = \text{const.}$ spacelike hypersurfaces, i.e. Coriolis-type effects are absent in this frame.

(d) The travel times for light rays (photons) circling the black hole in the prograde and retrograde directions at fixed (r,θ) are equal if measured in the LNRF.

(d) The four-velocity in this frame has the LNRF components $u^{\hat{a}} = u^{b}\, e^{\hat{a}}_{\ b}$, where the u^{b} can be calculated with Eq. (3.22). The LNRF observer has four-velocity $u^{\hat{t}} = 1$, $u^{\hat{i}} = 0$. The three-velocity relative to the LNRF has the components

$$v^{\hat{i}} = \frac{u^a e^{\hat{i}}{}_a}{u^b e^{\hat{t}}{}_b} \quad , \qquad i = r, \theta, \varphi. \tag{3.38}$$

For circular, equatorial orbits the only non-zero component is

$$\vartheta \equiv v^{\hat{\varphi}} = \frac{u^{\hat{\varphi}}}{u^{\hat{t}}} = \pm \frac{(r^2 \mp 2aM^{1/2}r^{1/2} + a^2)M^{1/2}}{(r^{3/2} \pm aM^{1/2})\ \Delta_K^{1/2}} \quad . \tag{3.39}$$

In the LNRF, a particle's four-momentum has then the same form as in flat spacetime, namely

$$p = \mu\gamma(1,\mathbf{v}) \quad , \quad \gamma = (1 - \mathbf{v}\cdot\mathbf{v})^{-1/2} \quad . \tag{3.40}$$

(f) The LNRF is not geodesic. In the Schwarzschild spacetime it corresponds to the static frame at constant (t,r,θ,φ).

3.5 Discussion

We now turn back to the question: when is a particle relativistic? In previous sections it was seen that $r_o \to r_\gamma$ implied $\gamma \gg 1$. This condition is identical with the flat space condition at spatial infinity where $\gamma = (1 - v^2)^{-1/2}$. However, as pointed out by Chrzanowski (1973), this is not at all satisfactory and should rather be replaced by some local criterion; the particle might not necessarily radiate when its energy with respect to infinity is large, or it might radiate even when γ is small. Secondly, the conserved parameter γ is not always a measure of the particle's energy, as is indeed the case in the ergosphere of the Kerr metric.

The measure of the particle's local energy is its time component $p_{\hat{t}}$ of the four-momentum p_a evaluated in the LNRF. This quantity is given by

$$p_{\hat{t}} = p_a\, e^a{}_{\hat{t}} = \mu\left[\frac{B}{\Sigma\,\Delta_K}\right]^{1/2}\left[\gamma - \omega_B \ell_z\right] \equiv \mu\,\gamma_B\,. \tag{3.41}$$

Hence, for the Kerr geometry the condition $\gamma \gg 1$ should be replaced by $\gamma_B = p_{\hat{t}}/\mu \gg 1$. On the other hand, for energetic orbits $r_o \simeq r_\gamma$ one obtains from Eq. (3.22) that $\gamma_B \gg 1$ whenever $\gamma \gg 1$. This again excludes stable circular orbits as possible sources of GSR. The condition $\gamma_B \gg 1$ also holds in the Schwarzschild case $(a = 0)$, where

$$p_{\hat{t}} = (1 - 2M/r_o)^{-1}\mu\gamma , \tag{3.42}$$

and $r_o \gtrsim 3M$. The synchrotron model therefore employs as sources particles which are aimed carefully at the black hole with an appropriate impact parameter as given in (3.33). They then achieve the unstable high-energy circular orbit close to the circular photon orbit.

There the particle is able to spiral many times until it either escapes again to infinity or until radiation reaction perturbs the orbit into a spiral- or dive-in orbit.

The situation differs though for particles in plunge orbits; i.e. when they drop off the last stable circular orbit - a situation likely to occur in astrophysical circumstances. Furthermore, in addition the test particle approximation will be made. Let q be the test particle's charge and μf be the coupling of the particle to the respective field. Then

$$\mu/M \ll 1, \qquad \mu f/M \ll 1, \qquad q/M \ll 1 \tag{3.43}$$

are the restrictions, which make the geodesic motion a good representation for the actual orbit. It also allows us to neglect the stress-energy of the particle's field in Einstein's equations; ie. there is no feedback on the background geometry.

IV. ORDINARY SYNCHROTRON RADIATION

4.1 Introduction

One of our goals is to derive properties of radiation emitted by particles moving on geodesics. It is therefore of interest for several reasons to consider first a simpler problem, viz. an accelerated particle in flat spacetime. This will serve especially as an introductory exercise for the subsequent calculations of GSR. It also will allow us to exhibit the main differences between the flat and curved space calculations.

Standard theory of synchrotron radiation considers charges accelerated in electromagnetic fields and emitting electromagnetic waves. Instead of repeating this wellknown calculation here, the analogous problem with scalar radiation will be solved in detail within a theory which is similar to Nordstrøm's theory. These results will be compared both with electromagnetic and gravitational synchrotron radiation from accelerated particles and that from geodesically moving particles.

The reason for considering scalar radiation at all - no massless scalar field has yet been found in Nature - lies mainly in the simplicity of the calculations. Only scalar spherical harmonics are needed for separation of variables in the corresponding wave equation. Also, in certain situations the higher spin wave equations give results similar to those obtained from scalar wave equations; i.e., in fact, certain properties will turn out to be spin-independent. In addition, calculations in flat spacetime generally illustrate the methods employed later without introducing the complications due to spacetime curvature.

4.2. Scalar Synchrotron Radiation in Flat Spacetime

Within the theory of Special Relativity, a massless scalar field ϕ is coupled to a particle moving on the world-line $z^a(\tau)$ through the action

$$I = -\frac{1}{8\pi}\int d^4x\ \eta^{ab}\phi_{,a}\phi_{,b} - \mu\int d\tau(1+f\phi)\left[-\eta_{ab}\frac{dz^a}{d\tau}\frac{dz^b}{d\tau}\right]^{1/2} . \qquad (4.1)$$

Here μ is the mass of the particle and η^{ab} the Minkowski metric. The coupling constant f leads to the Newtonian gravitational interaction in the static case if $f = \sqrt{G}$, where G is the Newtonian constant of gravity (otherwise set to unity here). Variation of ϕ in Eq. (4.1) leads to the inhomogeneous wave equation

$$\Box\phi = 4\pi f T . \qquad (4.2)$$

The scalar source T is the trace of the particle's canonical stress-energy tensor,

$$\begin{aligned} T &= \mu\int d\tau\ \eta_{ab}\ u^a u^b \delta^4[x-z(\tau)] \\ &= -\mu\,\gamma^{-1}\delta^3[\underline{x}-\underline{z}(\tau)] , \end{aligned} \qquad (4.3)$$

where $u^a = dz^a/d\tau$, $\gamma = dt/d\tau$. The technique now used to obtain the angular distribution of radiation is analogous to that in, e.g., Jackson (1967) for electromagnetic radiation. One does not wish to consider here the radiation reaction (as given implicitely through the action I) but rather assumes that the motion of the particle is determined by some other force which shall not be specified (example: the particle is held by a string). In the following the radiation is calculated as produced by a particle on a prescribed orbit. It is assumed without further analysis that the physically relevant solution of (4.2) is the retarded one.

Angular Distribution of Radiation

The solution of Eq. (4.2) in terms of a retarded Green's function is

$$\begin{aligned} \phi(x^a) &= \int d^4x'\ G(x^a,x'^a) f\,T(x'^a) \\ &= -2f\mu\int dt'\ \frac{\delta(t'+R(t')-t)}{\gamma R(t')} , \end{aligned} \qquad (4.4)$$

where $\underline{R}(t') = [\underline{x} - \underline{x}'(t)]$ is the vector connecting particle position and observer, i.e., we identify the primed coordinates with the particle z^a, the unprimed coordinates with the observer. Define the instantaneous 3-dimensional coordinate velocity $\underline{v} = d\underline{x}/dt'$ and use

$$\delta(t' + R(t') - t) = \delta(t' - t)(1 - \underline{R}\cdot\underline{v}/R)^{-1}_{ret}$$

(ret = evaluated at time t'); then Eq. (4.4) gives

$$\phi(x^a) = -\left[\frac{f\mu}{\gamma(R - \underline{R}\cdot\underline{v})}\right]_{ret} \quad . \tag{4.5}$$

One obtains the stress-energy tensor of the scalar field with positive-definite energy density from Eq. (4.1), with $L = \frac{1}{8\pi}\phi_{,a}\phi^{,a}$, as

$$T_{ab} = \frac{\partial L}{\partial(\phi^{,a})}\phi_{,b} - \eta_{ab}L = \frac{1}{4\pi}(\phi_{,a}\phi_{,b} - \frac{1}{2}\eta_{ab}\phi_{,c}\phi^{,c}) \quad .$$

Then $T_{oo} \equiv \varepsilon = \frac{1}{4\pi}[\dot{\phi}^2 + (\nabla\phi)^2] \geqslant 0$. One may define the Poynting vector $\underline{S}$ of the scalar field as

$$S^k \equiv T^{ok} = \frac{-1}{4\pi}\phi_{,o}\phi_{,k} \quad (k = 1,2,3) \quad . \tag{4.6}$$

The Poynting vector satisfies a continuity equation $\underline{\nabla}\,\underline{S} + \dot{\varepsilon} = 0$. Use Eq. (4.5) to calculate $\phi_{,o}$ and $\phi_{,k}$ and keep only terms of order R^{-1} for the radiation field to get

$$\phi_{,o} = -f\mu\left[\frac{(\underline{R}\cdot\dot{\underline{v}})R}{\gamma(R - \underline{R}\cdot\underline{v})^3} + \frac{(\underline{v}\cdot\dot{\underline{v}})R}{(\underline{R} - \underline{R}\cdot\underline{v})^2}\right]_{ret} \quad ,$$

$$\phi_{,k} = f\mu\left[\frac{(\underline{R}\cdot\underline{v})R_k}{\gamma(\underline{R} - \underline{R}\cdot\underline{v})^3}\right]_{ret} \quad . \tag{4.7}$$

In order to obtain (4.7) we made use of the relations

$$\frac{\partial t_{ret}}{\partial t} = \left[\frac{R}{\underline{R} - \underline{R}\cdot\underline{v}}\right]_{ret}, \quad \frac{dR(t)}{dt} = -\frac{\underline{R}\cdot\underline{v}}{R}, \quad (\underline{R}\cdot\underline{v})_{,k} = v_k + (\underline{R}\cdot\dot{\underline{v}} - v^2)\frac{\partial t_{ret}}{\partial t} \quad .$$

The radiated power per unit solid angle is given by

$$\frac{dP^{(0)}(t')}{d\Omega} = R^2(\underline{S}\cdot\underline{n})\frac{\partial t_{ret}}{\partial t} = R(\underline{S}\cdot\underline{R})\frac{\partial t_{ret}}{\partial t} \quad ,$$

($\underline{n}$ is the unit vector in the direction of $\underline{R}$, i.e., $\underline{n} \equiv \underline{R}/R$) as in

Eq. (4.7) $\underline{S} \sim \underline{R}$. The upper index indicates $s=0$ for scalar radiation. Inserting expressions (4.7) into the power formula, one obtains

$$\frac{dP^{(0)}}{d\Omega} = - f_\mu^2 {}^2 \dot{v}^2 \left[\frac{\cos^2(\theta+\alpha)}{\gamma^2(1 - v\cos\theta)^5} + \frac{v\cos(\theta+\alpha)\cos\alpha}{(1 - v\cos\theta)^4} \right] , \qquad (4.8)$$

where the angles α and θ are depicted below:

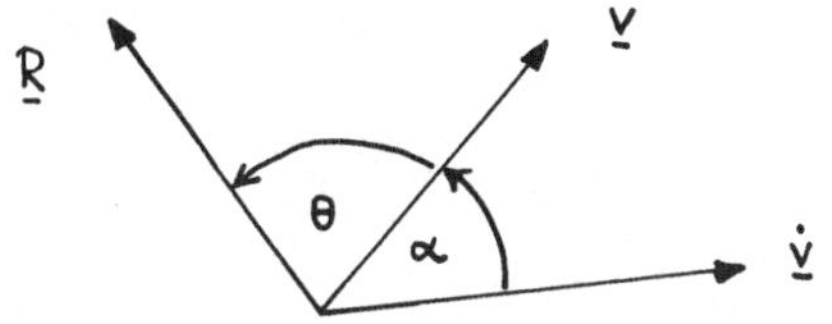

Now we specialize to circular motion ($\underline{v}\cdot\dot{\underline{v}} = 0$ or $\alpha = \frac{\pi}{2}$):

$$\frac{dP^{(0)}}{d\Omega} = - f_\mu^2 {}^2 \dot{v}^2 \gamma^{-2} \frac{\sin^2\theta}{(1 - v\cos\theta)^5} \quad . \qquad (4.9a)$$

For collinear acceleration ($\underline{v}\cdot\dot{\underline{v}} = v\cdot\dot{v}$ or $\alpha = 0$) one obtains:

$$\frac{dP^{(0)}}{d\Omega} = - f_\mu^2 {}^2 \dot{v}^2 \left[\frac{\cos^2\theta}{\gamma^2(1 - v\cos\theta)^5} + \frac{v\cos\theta}{(1 - v\cos\theta)^4} \right] \quad . \qquad (4.9b)$$

Both angular distributions are plotted in Fig. 3. The scalar radiation from circular source motion is - up to an intensity factor γ^{-2} - identical to electromagnetic synchrotron radiation from charges in linear motion ($\underline{v} \parallel \dot{\underline{v}}$) (see Sec. 4.3). The halfwidth-angle of the radiation becomes in the relativistic case ($v \to 1$) $\Delta\vartheta \sim \gamma^{-1}$ ($\vartheta \equiv \pi/2 - \theta$).

Spectrum of Radiation

Next the frequency distribution of scalar synchrotron radiation has to be found (Moncrief 1972, Chrzanowski 1973), where a technique will be used which differs from the retarded Green's function method. This new method has the advantage that it can be employed later for wave equations in black hole geometries. Eq. (4.2) shall be solved by separation of variables from which one can construct a Green's function for the radial part. This Green's function will be composed of two specific solutions of the homogeneous equation which are matched together at the source and which satisfy certain boundary condition at infinity. It is assumed that the general solution of $\Box\phi = 0$ can be

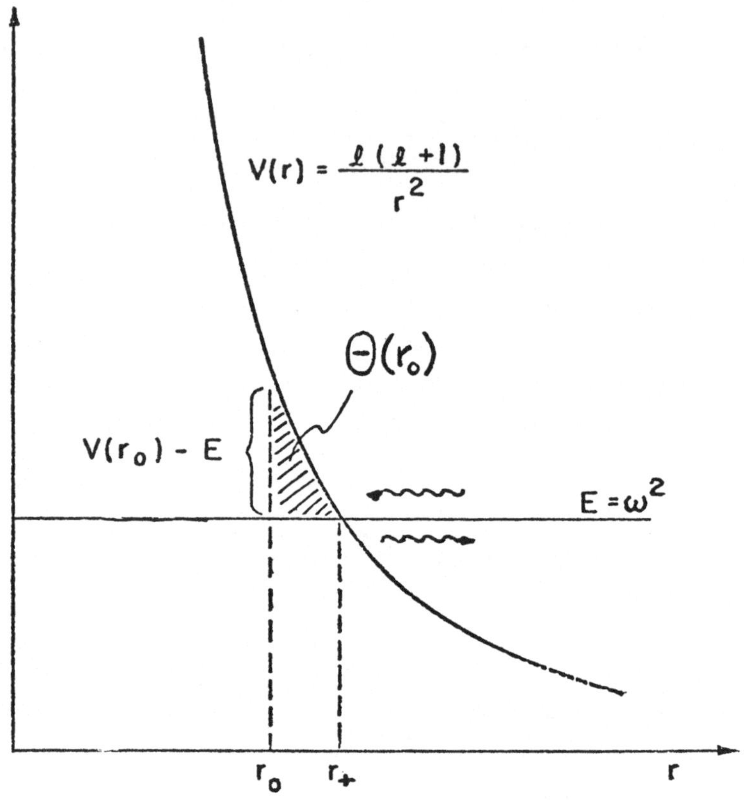

Figure 3. Effective potential of scalar radial wave equation in flat spacetime. The homogeneous solution $u^{reg}_{\ell m\omega}(r)$ can be thought of as a wave of energy $E = \omega^2$ incident from infinity and reflected at the effective potential $V(r)$. The hatched area is a measure for the penetration barrier $\theta(r_o)$. If it is large, the WKB approximation is valid. Synchrotron radiation occurs if only if $\theta(r_o) \ll 1$, i.e. if $r_o \simeq r_+$. There the WKB approximation breaks down and a different method has to be used to find $u^{reg}_{\ell m\omega}(r_o)$.

expressed in the form of a Fourier expansion

$$\phi(t,r,\theta,\varphi) = \sum\!\!\!\!\!\!\int \phi_{\ell m\omega}(r)\; Y_{\ell m}(\theta,\varphi)\, e^{-i\omega t} , \qquad (4.10)$$

where $\sum\!\!\!\!\!\!\int \equiv \sum_{\ell=0}^{\infty} \sum_{m=-\ell}^{\ell} \int_{-\infty}^{\infty} d\omega .$

The angular functions $Y_{\ell m}(\theta,\varphi)$ are the usual scalar spherical harmonics. The radial Fourier-component $\phi_{\ell m\omega}(r)$ satisfies the radial equation

$$\left[\frac{d^2}{dr^2} + \frac{2}{r}\frac{d}{dr} - \frac{\ell(\ell+1)}{r^2} + \omega^2 \right] \phi_{\ell m\omega}(r) = 0 . \qquad (4.11)$$

The substitution $u_{\ell m\omega}(r) = r\phi_{\ell m\omega}(r)$ reduces the inhomogeneous wave equation to

$$\left[\frac{d^2}{dr^2} + V(r) - \omega^2 \right] u_{\ell m\omega}(r) = 2fr\int d^4x' T\, \bar{Y}_{\ell m}(\theta',\varphi')\, e^{i\omega t'} , \qquad (4.12)$$

where $V(r) = \ell(\ell+1)/r^2$. In order to ensure that one obtains the retarded solution of (4.12), one demands as boundary conditions regularity at the origin and outgoing waves at infinity.

$$u_{\ell m\omega}(r) = 2f \int d^4x_o \sqrt{-\eta}\; T(x_o) G(r,r_o)\; \bar{Y}_{\ell m}(\theta_o,\varphi_o) . \qquad (4.13)$$

Here the Green's function $G(r,r_o)$ is a solution of

$$\left[-\frac{d^2}{dr^2} + V(r) - \omega^2 \right] G(r,r_o) = \delta(r - r_o) . \qquad (4.14)$$

Now two independent solutions of Eq. (4.12) will be obtained. The first u^{reg} is regular at the regular singular point $r = 0$ of Eq. (4.12). The second u^{out} is purely outgoing at the irregular singular point $r = \infty$. In the standard WKB-form they are

$$\begin{aligned} u^{reg} &= |\omega|^{-1/2} \left[e^{-i\omega r} + \bar{S}\, e^{i\omega r} \right] , \\ u^{out} &= |\omega|^{-1/2} e^{i\omega r} , \end{aligned} \qquad r \to +\infty , \qquad (4.15)$$

where S is a complex reflection amplitude. Then the Green's function incorporating these boundary conditions is given for each mode $\ell m\omega$ by

$$G(r,r_o) = \begin{cases} A\, u^{out}(r) = \frac{A}{W}\, u^{reg}(r_o)\, u^{out}(r) , & r > r_o , \\ B\, u^{reg}(r) = \frac{B}{W}\, u^{out}(r_o)\, u^{reg}(r) , & r < r_o , \end{cases} \qquad (4.16a)$$

where W is the Wronskian of the two homogeneous solutions. The Wrons-

kian is independent of the variable r and may therefore be easily evaluated at $r \to \infty$. The constants A and B in Eq. (4.16) are fixed by the matching conditions of continuity and a discontinous derivative at $r = r_o$, i.e. $[G] = 0, \left[\frac{dG}{dr}\right] = -1$ at $r = r_o$.
Thus

$$G(r,r_o) = \frac{i}{2}\frac{\omega}{|\omega|}\begin{cases} u^{reg}_{\ell m\omega}(r_o)\; u^{out}_{\ell m\omega}(r) & , \quad r > r_o \\ u^{out}_{\ell m\omega}(r_o)\; u^{reg}_{\ell m\omega}(r) & , \quad r < r_o \quad . \end{cases} \tag{4.16b}$$

Then, by using Eq. (4.12) one finds for the $r > r_o$ part of the solution

$$u_{\ell m\omega}(r) = if\, u^{out}_{\ell m\omega}(r) \int d^4x_o \sqrt{-\eta}\; T(x_o)\; \bar{Y}_{\ell m}(\theta_o, \varphi_o)\; u^{reg}_{\ell m\omega}(r_o)\; e^{i\omega t_o} . \tag{4.17}$$

If we transform back to the original field by $\phi(r) = r^{-1}u(r)$, this can conveniently be written as

$$\phi(t,r) = \sum\!\!\!\!\!\!\int\; i\frac{\omega}{|\omega|}\; \phi^{out}_{\ell m} \;\langle \phi^{reg}_{\ell m}\, ,\, fT \rangle \; . \tag{4.18}$$

Here

$$\phi^{reg}_{\ell m\omega}(r) = \frac{\bar{u}^{reg}_{\ell m\omega}}{r}\, Y_{\ell m}(\theta,\varphi)\; e^{-i\omega t} \;,$$

$$\phi^{out}_{\ell m\omega}(r) = \frac{u^{out}_{\ell m\omega}}{r}\, Y_{\ell m}(\theta,\varphi)\; e^{-i\omega t} \;, \tag{4.19}$$

are solutions of Eq. (4.11) and the inner product is defined as

$$\langle \phi, T \rangle = \int d^4x \sqrt{-\eta}\; \bar{\phi}(x)\, T(x) \; . \tag{4.20}$$

The amplitude of each mode $\phi^{out}_{\ell m\omega}$ in (4.18) depends on how strongly an associated wave state $\phi^{reg}_{\ell m\omega}$ couples to the source of the radiation.

The total energy radiation to future null infinity, $\mathscr{I}^+$, is

$$E^{(o)} = -\lim_{r\to\infty} \int d\Omega\, r^2\, T^r_t = -\lim_{r\to\infty} \int d\Omega \frac{r^2}{4\pi}\, \phi^{,r}\phi_{,t} \equiv \int d\omega \frac{dE^{(o)}}{d\omega} . \tag{4.21}$$

Asymptotically, $u^{out} \sim |\omega|^{-1/2} \exp(i\omega r)$ in (4.19) and hence

$$\phi(t,r) = \oint i \frac{\omega}{|\omega|^{3/2}} \langle \phi^{reg}_{\ell m}, fT \rangle Y_{\ell m}(\theta,\varphi)\, r^{-1} e^{-i\omega(t-r)} . \qquad (4.22)$$

This yields with Eq. (4.21)

$$\frac{dE^{(o)}}{d\omega} = \sum_{\ell=0}^{\infty} \sum_{m=-\ell}^{\ell} \omega \left| \langle \phi^{reg}_{\ell m}, fT \rangle \right|^2 . \qquad (4.23)$$

When the source is a particle with scalar charge μ in circular orbit at $r = r_o$, $\theta = \pi/2$ with angular frequency ω_o, one gets with Eq. (4.3)

$$T = -\mu \gamma^{-1} r^{-2} \delta(r-r_o) \delta(\theta - \pi/2) \delta(\varphi - \omega_o t), \qquad (4.24)$$

where $\gamma = dz^o/d\tau$ is the energy per unit mass. The particle is relativistic when $\gamma \gg 1$ and the circular orbit condition implies $\omega = m\omega_o$. Then one inserts (4.24) into (4.23) to get

$$\frac{dE^{(o)}}{d\omega} = 4\pi^2 \sum_{\ell,m} \omega \left| u^{reg}_{\ell m}(r_o) \right|^2 \left| Y_{\ell m}(\pi/2,0) \right|^2 \delta(m\omega_o - \omega)^2 .$$

Note that $\delta(m\omega_o - \omega)^2 = \delta(0)\delta(m\omega_o - \omega)$. For stationary systems the power $P^{(o)}$ radiated is defined through $E^{(o)} = \int_{-T}^{T} P^{(o)} dt = 2\pi P^{(o)} \delta(0)$ for a system which remains stationary for a time $T \to \infty$. Then with $P^{(o)} = \int d\omega \frac{dP^{(o)}}{d\omega}$ the power per unit frequency is given by

$$\frac{dP^{(o)}}{d\omega} = 2\pi \left[\frac{f\mu}{r_o \gamma} \right]^2 \sum_{\ell,m} m \left| u^{reg}_{\ell m}(r_o) \right|^2 \left| Y_{\ell m}(\pi/2,0) \right|^2 \delta(\omega - m\omega_o) . \qquad (4.25)$$

Subsequently, Eq. (4.25) is evaluated explicitely.

WKB Approximation

In order to complete the derivation of the frequency spectrum, in Eq. (4.25) a solution $u^{reg}_{\ell m, m\omega_o}(r_o)$ of the homogeneous equation

$$-\frac{d^2u}{dr^2} + \left[V(r) - \omega^2 \right] u = 0 \qquad (4.26)$$

is required. Since our interest is in synchrotron radiation, a high-frequency approximation of the potential in (4.26) is sought. Define $k = \ell - m$ and note that $r_o \omega_o = [1-\gamma^2]^{1/2}$, then the potential can be rewritten as

$$\kappa^2 = V(r) - \omega^2 = \frac{l(l+1)}{r^2} - m^2\omega_o^2 = \frac{m^2+m(1+2q)+O(m^o)}{r^2} + \frac{m^2}{r_o^2}(1-\gamma^{-2}) . \tag{4.27}$$

A WKB solution of Eq. (4.26) is possible when k is a slowly varying function of position, i.e. when the condition

$$|\kappa^2|^{-1}\left|\frac{d\kappa}{dr}\right| = \frac{\frac{d}{dr}\left[V-\omega^2\right]}{2\left[V-\omega^2\right]^{3/2}} \gg 1 \tag{4.28}$$

is satisfied. Then the WKB approximation is valid, which allows one to write the solution in the form

$$u^{reg}_{\ell m, m\omega_o}(r_o) = \left[\kappa(r_o)\right]^{-1/2} \exp\left[-\Theta(r_o)\right] . \tag{4.29}$$

Here $\Theta(r_o)$ is the barrier penetration factor

$$\Theta(r_o) = \int_{r_o}^{r_+} \sqrt{V(r) - \omega^2}\, dr , \tag{4.30}$$

where r_+ is the classical turning point defined by $V(r_+) = \omega^2$ (see Fig. 3). Since the particle circles at less than the speed of light, the orbit radius r_o can be shown to be always under the barrier, i.e. $r_o < r_+$. In particular, when conditions (4.31b) are satisfied, r_o lies deep within the barrier and the WKB approximation is justified. Condition (4.28) holds whenever

$$m^{2/3}\gamma^{-2} + \frac{k}{2} m^{-1/3} + 1 \gg 1 , \tag{4.31a}$$

which is equivalent to

$$m \gg \gamma^3 \quad \text{or} \quad k \gg m^{1/3}. \tag{4.31b}$$

Now linearize $V - \omega^2$ around the classical turning point,

$$\Theta(r_o) \simeq -\frac{2}{3}\left[V(r_o)-\omega^2\right]^{3/2}\left[\frac{d(V-\omega^2)}{dr}\right]^{-1}\Bigg|_{r=r_o} \simeq \frac{1}{3}\left[\frac{Q}{m}\right]^{1/2} , \tag{4.32}$$

where $Q = 1 + 2k + m\gamma^{-2}$. Then

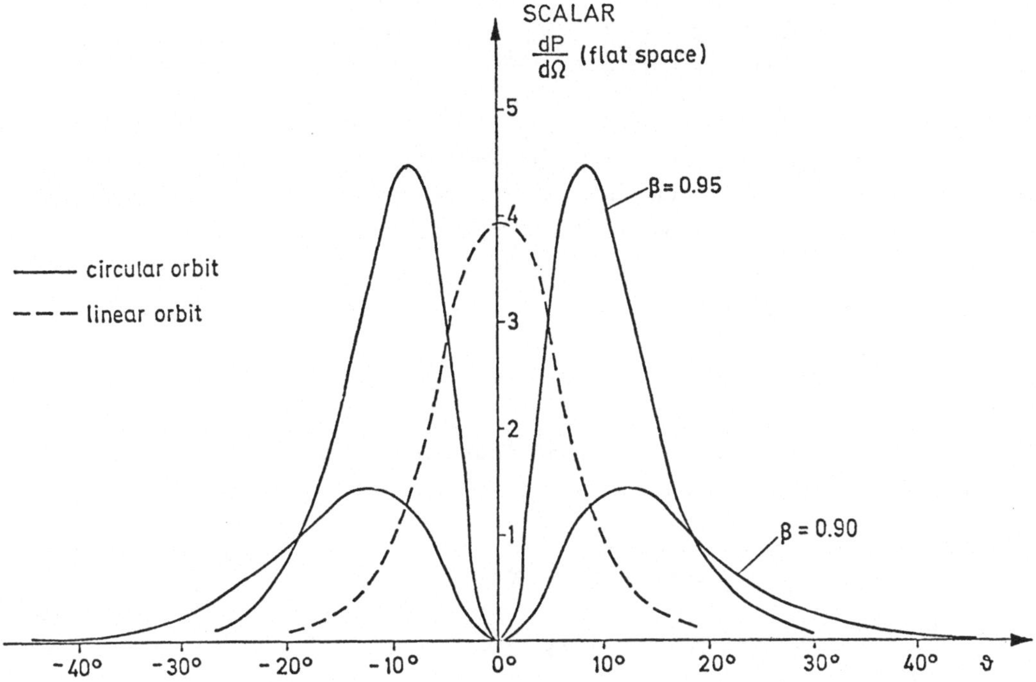

Figure 4. Scalar Synchrotron Radiation from Accelerated Particles in Flat Space

$$\left| u^{reg}_{\ell m, m\omega_o}(r_o) \right|^2 \simeq \frac{r_o}{[mQ]^{1/2}} \exp\left[-2\,\Theta(r_o)\right]$$

and with the help of Eq. (4.27)

$$P^{(o)} = 2\pi \left[\frac{f\mu}{r_o\gamma}\right]^2 \sum_{\ell,m} \left[\frac{m}{Q}\right]^{1/2} \left| Y_{\ell m}(\pi/2,0)\right|^2 \exp\left[-2\Theta(r_o)\right] . \qquad (4.33)$$

Expression (4.33) indicates that the factor $\Theta(r_o)$ governs the intensity of the radiation. Each time Θ increases by one, the power e-folds twice. This property can be used to truncate the summation over $k = \ell - m$ in Eq. (4.33). The k sum can be effectively cut off at a k_{crit} defined by

$$\Theta(k = k_{crit}, m) = \Theta(k = 0, m) + 1 , \qquad (4.34)$$

because terms with $q > q_{crit}$ contribute a factor of e^{-2} less then the leading terms. From (4.32) one obtains together with (4.34)

$$k_{crit} \simeq \begin{cases} m^{1/3} & \text{for} \quad m \ll \gamma^3 , \\ \gamma & \text{for} \quad m \gg \gamma^3 . \end{cases} \qquad (4.35)$$

One also wants to find the critical frequency $\omega_{crit} = m_{crit}\,\omega_o$ above which the spectrum is exponentially damped for any value of k. From the corresponding condition

$$\Theta(k = 0,\ m = m_{crit}) = \Theta(k = 0, m = 0) + 1 \qquad (4.36)$$

and Eq. (4.32) one has $m_{crit} = \gamma^3$. In summary the bulk of the energy is radiated into modes with

$$\omega \lesssim \omega_{crit} = m_{crit}\,\omega_o = \gamma^3 \omega_o, \quad k \lesssim k_{crit} = m^{1/3} \sim \gamma . \qquad (4.37a)$$

Another property characterizing the radiation is the latitudinal beaming in the relativistic case. From the discussion after Eq. (4.9) it was found that $\Delta\vartheta \sim \gamma^{-1}$. But at frequencies beyond the peak-intensities one has $\gamma \sim m^{1/3}_{crit}$ and hence

$$\Delta\vartheta \sim m^{-1/3} . \qquad (4.37b)$$

Relations (4.37) are the main properties of scalar radiation of interest in this context.

To obtain the spectrum consider the two cases $\omega \gg \omega_{crit}$ and $\omega \lesssim \omega_{crit}$. In the first case the WKB approximation is valid and Eq. (4.33) can be evaluated in the high-frequency-limit. The sum in this equation may be converted into a sum over k, which can be truncated at k_{crit} by condition (4.35). Also, with Stirling's formula, $|Y_{\ell=m+k,m}(\pi/2,0)|^2$ can be approximated. Then the spectrum becomes

$$\frac{dP^{(0)}}{d\omega} \sim \frac{f^2\mu^2}{r_o} \left[\frac{m}{\gamma}\right]^{1/2} \exp\left[-2m/3\gamma^2\right] \quad (\omega \gg \omega_o\gamma^3). \tag{4.38}$$

In the other case of the peak-frequencies ($\omega \lesssim \omega_{crit}$) a different method has to be used as the WKB form breaks down when $\theta(r_o) \ll 1$, i.e. $r_o \simeq r_+$ (see Fig. 3). Then (see, e.g., Schiff 1968) two linearly independent solutions to Eq. (4.14) are given by

$$u^{\pm}(r_o) \sim \left[\frac{\theta(r_o)}{k(r_o)}\right]^{1/4} I_{\pm\frac{1}{3}}\left[\theta(r_o)\right] , \tag{4.39}$$

where $I_{\pm\frac{1}{3}}$ are Bessel-functions of imaginary argument. Their asymptotic forms are

$$I_{\pm\frac{1}{3}}(\theta) \sim \begin{cases} [\theta/2]^{1/3} / \Gamma(1 \pm \frac{1}{3}) & , \quad \theta \rightarrow 0 \\ [2\pi\theta]^{-1/2}\left[e^{\theta} + e^{-\theta - i\pi(\frac{1}{2} \pm \frac{1}{3})}\right], & \theta \rightarrow \infty. \end{cases} \tag{4.40}$$

Solution u^{reg} is a linear combination of $u^{\pm}$ that assumes the above WKB form for $\theta \gg 1$. Generally, u^{reg} is expressed in terms of the Bessel functions as

$$u^{reg}(r_o) = \sqrt{\frac{2\pi}{3}} \left[\frac{\theta(r_o)}{k(r_o)}\right]^{1/4} \left[I_{-\frac{1}{3}}(\theta) - I_{\frac{1}{3}}(\theta) \right] , \tag{4.41}$$

which yields in the limit $\theta(r_o) \ll 1$

$$\left|u^{reg}(r_o)\right|^2 = \frac{2\pi\, 6^{2/3} r_o}{k m^{2/3}\left[\Gamma(2/3)\right]^2} . \tag{4.42}$$

Inserted into Eq. (4.25) this gives for the power spectrum

$$\frac{dP^{(0)}}{d\omega} \sim \int_0^{m^{1/3}} dk \, \frac{f^2\mu^2 m^{1/3}}{r_0\gamma^2} \left| Y_{m+k,m}(\pi/2,0) \right|^2 \sim \frac{f^2\mu^2}{r_0} m\,\gamma^{-2} \qquad (\omega \lesssim \omega_0\gamma^3)\,. \tag{4.43}$$

Although the scalar power spectrum differs from the analogous formuale for electromagnetic & gravitational synchrotron radiation it will be seen that properties (4.37) will hold in all cases. For comparison with radiation from particles moving on geodesics, see Table I.

4.3 Electromagnetic & Gravitational Synchrotron Radiation in Flat Spacetime

By the methods used in the previous section, the angular distribution of electromagnetic synchrotron radiation may be obtained (see, e.g. Jackson 1967). If a charge q is moving relativistically in linear motion ($\underline{v} \parallel \dot{\underline{v}}$) then the result is

$$\frac{dP^{(1)}}{d\omega} \sim q^2\dot{v}^2 \frac{\sin^2\theta}{(1-v\cos\theta)^5} \tag{4.44}$$

with a beaming angle of $\Delta\vartheta = \gamma^{-1}$ for $v \to c$. The index stands for the spin of the field $s=1$. For a charge in instantaneous circular motion ($\underline{v} \perp \dot{\underline{v}}$) one gets

$$\frac{dP^{(1)}}{d\omega} \sim q^2\dot{v}^2 \frac{1}{(1-v\cos\theta)^3}\left[1 - \frac{\sin^2\theta\cos^2\varphi}{\gamma^2(1-v\cos\theta)^2}\right], \tag{4.45}$$

where θ, φ are the usual angles defining the direction of observation. The total power found by integrating over (4.45) is then

$$P^{(1)} \simeq \frac{2}{3} q^2 \dot{v}^2 \gamma^4 \,. \tag{4.46}$$

Similarly, one may obtain that the spectrum is given by

$$\frac{dP^{(1)}}{d\omega} \sim q^2 \gamma \left[\frac{\omega}{\omega_{crit}}\right]^{1/2} \exp\left[-2\omega/\omega_{crit}\right], \tag{4.47}$$

where, as in the scalar case, $\omega_{crit} = \omega_0\gamma^3$.

The analogous situation has also been studied for gravitational radiation emitted by particles moving on accelerated orbits in flat spacetime (Price et al. 1973, Doroshkevich et al. 1973). In particular, a relativistic circular orbit is considered on which the particle is held by a thin rod. In this case the frequency spectrum is given by

$$\frac{dP^{(2)}}{d\omega} \sim \mu^2 \omega_o^2 \gamma \left[\frac{\omega_{crit}}{\omega}\right]^{1/3} . \tag{4.48}$$

Above the critical cut-off frequency ω_{crit} the spectrum falls off as $\exp(-2\,\omega/\omega_{crit})$. Incidently, the contribution of the rod to the power holding the mass has, to some extent, also been taken into account. For $\gamma = 100$, its contribution is 10^{-1} times the contribution of the particle (modulo the idealizations assumed for a thin, unbreakable rod). The total power of gravitational radiation is of the form

$$P^{(2)} \simeq \int d\omega \frac{dP^{(2)}}{d\omega} \sim \mu^2 \omega_o^2 \gamma^4 . \tag{4.49}$$

The critical frequency of exponential cut-off and the angle of the beam half-width have been obtained as

$$\omega_{crit} = \omega_o \gamma^3 , \qquad \Delta\vartheta = m^{-1/3} . \tag{4.50}$$

4.4 Discussion

The spectra of scalar (s = 0), electromagnetic (s = 1) and gravitational (s = 2) synchrotron radiation in flat spacetime may be summarized as

$$\frac{dP^{(s)}}{d\omega} \sim \gamma \omega_o^2 \left[\frac{\omega}{\omega_{crit}}\right]^{1-2s/3} \exp\left[-2m/3\gamma^3\right] \qquad (\omega \lesssim \omega_{crit}) . \tag{4.51}$$

In each case, the respective particle moves on an accelerated, relativistic circular orbit. Although the spectra differ for different spin, the characteristic properties for ω_{crit}, q_{crit} and beaming angle $\Delta\vartheta$ hold in all three cases. The main properties of synchrotron radiation in flat spacetime are listed in Table I. Although derived later (Chap. VIII), these features are contrasted with the analogous ones, which hold for synchrotron radiation as emitted by particles moving on geodesics in curved spacetime.

Table I. Comparison of synchrotron radiation from accelerated particles in flat spacetime to that from particles moving on a geodesic in curved spacetime.

		geodesic	accelerated
m_{crit}		$\sim \gamma^2$	$\sim \gamma^3$
k_{crit}	$m < m_{crit}$	~ 1	$\sim m^{1/3}$
	$m > m_{crit}$	~ 1	$\sim \gamma$
$\Delta\vartheta$	$m < m_{crit}$	$\sim m^{-1/2}$	$\sim m^{-1/3}$
	$m \simeq m_{crit}$	$\sim \gamma^{-1}$	$\sim \gamma^{-1}$
	$m > m_{crit}$	$\sim m^{-1/2}$	$\sim \gamma m^{-1/2}$
P		$\sim \frac{\mu^2}{M^2} \gamma^2$	$\sim \mu^2 \gamma^4$
$\frac{dP^{(s)}}{d\omega}$		Chapter IX	$\mu^2 \left[\frac{\omega}{\omega_{crit}}\right]^{1-2s/3} \times \exp\left[-2m/3\gamma^3\right]$

V. PERTURBATIONS OF SPACETIMES

5.1 Introduction

Perturbation methods in the realm of gravitational radiation theory have been most extensively developed to deal with certain problems of astrophysical interest. These methods deal with linearized equations and provide (hopefully) approximate solutions of nonlinear problems. Naturally, such a formal linearization procedure throws away all possible nonlinear effects hidden in the Einstein field equations. It is also clear that with a perturbation analysis only a limited class of problems can be adequately treated, which, although interesting, may quite often not be the really important problems. Although perturbation analysis is usually quite helpful, it may not be in some cases. For example, if the linearized treatment reveals instabilities, there is no guarantee that the system remains unstable when nonlinearities are included (see examples in Hydrodynamics!). One usually assumes that linearized stability implies stability. This is probably true (it is true for ordinary differential equations in finite dimensions by L'yapunovs theorem). Furthermore, it is not even known whether a solution of the linearized equations is always a linearized of some exact equations. This is the socalled question of "linearization stability". Brill & Deser (1973) have given examples of vacuum spacetimes which are linearization unstable. Fisher & Marsden (1973) as well as Choquet-Bruhat & Deser (1973) have shown that flat spacetime is linearization stable in this sense. Also, asymptotically flat spacetimes are probably Fischer-Marsen linearization stable. More general cases are the subject of more recent investigations (O'Murchada & York 1974; Moncrief 1975).

So far extensive studies have been made of black hole stability against infinitesimal perturbations by various kinds of fields. The evidence mounts that even rotating black holes are stable classically under small perturbations. Stability of nonrotating black holes was shown by Vishveshwara 1970; the corresponding perturbation equations for the rotating black hole have first been derived by Teukolsky 1972, 1974, and studied numerically by Press & Teukolsky 1974; for an almost analytical analysis see Stewart 1975.

However, very little is known so far about stability when the black hole is subjected to arbitrary, i.e. nonlinear, interactions with its environment. Press (1972) has studied the behaviour of black holes when immersed in a static scalar field. DeWitt et al. are investigating black hole - black hole collisions numerically. Also Hawking's first law of black hole thermodynamics should be mentioned in this connection. (Les Houches Lectures 1973). In addition, Hawking (1974) has shown that that small black holes with mass $\lesssim 10^{15}g$ may possibly decay due to pair production effects in strong gravitational fields. This constitutes, in a sense, a quantum-mechanical instability of black holes in general. Usually the difficulty of solving nonlinear differential equations restricts one automatically to the linearized perturbation approch, but for gravitational interactions the nonlinearities often contain those features that are of astrophysical and cosmological importance (an example is furnished by the problem of colliding gravitational waves; see Khan & Penrose, Nature, **229**, 185 (1971)).

Basically two approaches to the linearized treatment of gravitational perturbations have been developed. First, the standard linearized theory of the gravitational field in flat spacetime (see e.g. Landau-Lifshitz 1951, Chap. II, Trautmann 1962, Pirani 1964) and the Regge-Wheeler formalism deal with perturbations of the Schwarzschild metric itself (Regge & Wheeler 1957, Zerilli 1970, Vishveshwara 1971). Second, using the Newman-Penrose (NP-) formalism (Newman and Penrose 1965) one perturbs the tetrad components of the connection, the Weyl-curvature tensor, etc.

5.2 The Linearized Theory of Gravity

In this approach to perturbation theory one considers linearized perturbations h_{ab} of the Minkowski metric $(\eta_{ab}) = \mathrm{diag}(-1,+1,+1,+1)$, i.e. one puts

$$g_{ab} = \eta_{ab} + \varepsilon h_{ab} \ . \tag{5.1}$$

where ε is a magnitude-scaling parameter. Eq. (5.1) holds in any pseudo-Cartesian coordinate system $\{x^a\}$ on $\mathbb{R}^4$. All quantities derived from g_{ab} are expanded up to first order in ε. The quantities h_{ab}

are coordinate dependent. One can easily derive that under a coordinate transformation $\tilde{x}^a = x^a + \varepsilon \xi^a$, with ξ^a an arbitrary vector field, h_{ab} transforms as

$$h_{ab} \rightarrow \tilde{h}_{ab} = h_{ab} - 2 \nabla_{(b} \xi_{a)} . \qquad (5.2)$$

Transformations (5.2) are called coordinate gauge (CG) transformations. From h_{ab} one can derive the linearized curvature tensor $R_{abcd} = 2\varepsilon \nabla_{[d} h_{c][a;b]} = 2\varepsilon \partial_{[d} h_{c][a,b]}$ and its contractions. These turn out to be CG independent, i.e. unchanged up to first order in ε.

The CG dependence of the metric naturally carries over to metric perturbations. The use of CG dependent perturbations, however, only leads to difficulties if one asks the wrong questions. E.g., it is not meaningful to ask whether the density contrast $\delta\varrho/\varrho$ in cosmology is large or small at a particular event (x^a), but it is all right to ask what is the total mass loss $(\delta M)_{total}$ in a bounded disturbance.

5.3 Definition of Perturbation

Here we have to specify what we want to call a perturbation. We shall see that all ambiguities are avoided if only gauge independent perturbations are considered, but this cannot always be done. A satisfactory characterization of perturbations in general has not been given. For reviews see Weinberg (1972); Sachs (1973). We essentially follow the construction given by Stewart and Walker (1974).

Let a spacetime be a pair (M,g) consisting of a smooth, four-dimensional, connected Hausdorff manifold M, and a smooth Lorentz metric g of signature (-+++) defined on M. Consider an unperturbed spacetime (M,g) and a perturbed spacetime (M',g'), slightly different from (M,g) because of the perturbation. Both M and M' have the same topology and differentiable structure and it is assumed that the perturbation is continuous in the sense that both spacetimes are connected by a path in the space of spacetimes. Parametrize this family of spacetimes as $(M_\varepsilon, g(\varepsilon))$ so that one has (M,g) for $\varepsilon = 0$ and (M',g') for $\varepsilon = 1$. Following Geroch (1969) one requires that $(M_\varepsilon, g(\varepsilon))$ satisfy:

P1: There exists a smooth, Hausdorff, five-dimensional manifold $\mathcal{M}$ in which the M_ε are smooth, properly imbedded, non-intersecting four-dimensional submanifolds.

P2: The $g(\varepsilon)$ define a smooth tensor field $g^{\alpha\beta}$ $(\alpha, \beta = 1,\dots,5)$ of signature $(0 - + + +)$ on $\mathcal{M}$. Specifically, $g^{\alpha\beta}\varepsilon_{,\beta} = 0$; i.e. the singular hypersurfaces $g^{\alpha\beta}(d\varepsilon)_\beta = 0$ of $g^{\alpha\beta}$ are the M_ε, and $g^{\alpha\beta}$ induces the metric $g^{ab}(\varepsilon)$ on each M_ε.

The reason for assigning signature 0 to the ε-dimension of $\mathcal{M}$ is to avoid questions like "What is the distance between two spacetimes?"

Now in order to relate points on M_o and M_ε we need a diffeomorphism $\phi_\varepsilon : M_o \to M_\varepsilon$ which defines when a point $\phi_\varepsilon(P_o) \equiv P_\varepsilon$ is to be regarded as "the same point" as $P_o \in M_o$. This is called an <u>identification map</u>. The map ϕ_ε is not uique, its arbitrariness corresponds exactly to gauge transformations; it can be described as follows: consider any smooth, nowhere-vanishing differentiably vector field V on $\mathcal{M}$ which is transverse (nowhere tangent) to the M_ε. The vector field V induces a local one-parameter group of diffeomorphisms ϕ_ε on $\mathcal{M}$ (Trautmann 1964, p. 88, or Hawking & Ellis 1973) and we identify points P_o and P_ε when they both lie on the same orbit generated by this group.

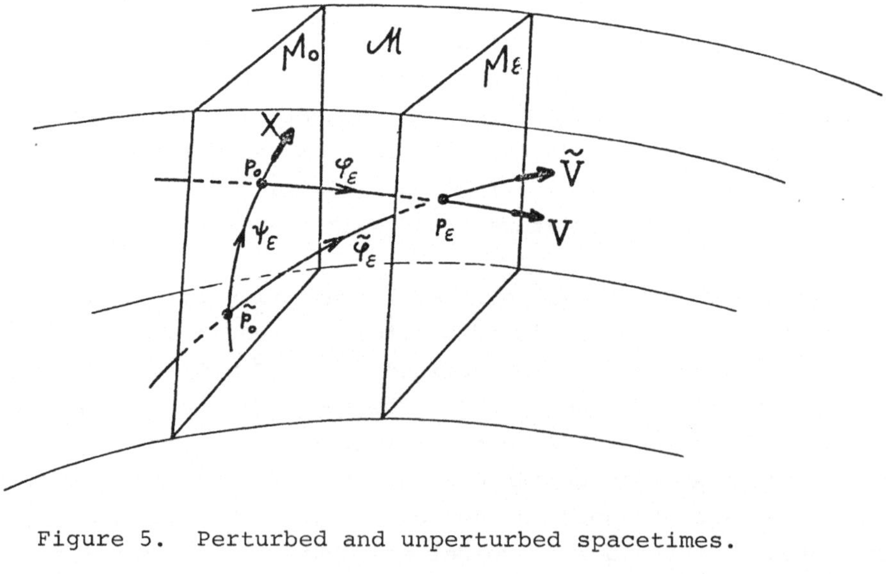

Figure 5. Perturbed and unperturbed spacetimes.

Considering a field of linear geometric objects Q_o on M_o and Q_ε on M_ε, we have a field of quantities Q on $\mathcal{M}$. The <u>linearization</u> Q_1 of Q_ε whith respect to the vector field V is the Lie-derivative of Q in the direction of V evaluated at the unperturbed spacetime M_o :

$$Q_1 = \mathcal{L}_V Q \Big|_{\mathcal{M}_o} \tag{5.3}$$

Thus we have on M_ε :

$$Q_\varepsilon = \phi_{\varepsilon *}(Q_o + \varepsilon Q_1) + O(\varepsilon^2) . \tag{5.4a}$$

Equivalently, the perturbation can be expressed on M_o as

$$Q_1 = \lim_{\varepsilon \to 0} \frac{1}{\varepsilon}\left[\phi_{\varepsilon *}^{-1}(Q_\varepsilon) - Q_o \right] . \tag{5.4b}$$

The difference of two perturbations Q_1 and $\tilde{Q}_1$ due to different choices V and $\tilde{V}$ is then given by slight generalization of a lemma of Sachs (1964):

Lemma: Let $(M_\varepsilon, g^{ab}(\varepsilon))$ be a one-parameter family of spacetimes constructed as above and satisfying conditions P1 and P2. Let $\mathcal{M}$ be the five-manifold and Q a field of linear geometrical objects on $\mathcal{M}$. Let V and $\tilde{V}$ be two vector fields on $\mathcal{M}$ transverse to the M_ε. Then the corresponding linearized perturbations Q_1 and $\tilde{Q}_1$ of Q_o are related by

$$Q_1 - \tilde{Q}_1 = \mathcal{L}_X Q_o , \tag{5.5}$$

where $X = V - \tilde{V}$ is a vector field tangent to M_o.

This follows immediately from (5.3).

Example: In the linearized theory of gravity the transformation $x^a \to x^a + \varepsilon \xi^a$ is a change of vector fields $V \to \tilde{V}$; hence the resulting change in the perturbation h_{ab} as given by Eq. (5.2) can be written as

$$h_{ab} - \tilde{h}_{ab} = 2 \nabla_{(b} \xi_{a)} = \mathcal{L}_\xi \eta .$$

Although in the example just given the gauge transformation was given in a specific coordinate system, coordinates are not essential for the question of gauge. A certain gauge is fixed by the choice of identification map which relates points of the unperturbed to those of the perturbed manifold; for this no coordinates have to be introduced. Such a gauge, formulated in coordinate-free manner, may therefore be called a point gauge (PG). Only when coordinates are employed is the point gauge a coordinate gauge. In order to avoid the above-mentioned difficulties with PG dependent perturbations we single out quantities

Q_1 which are independent of the specific choice of identification map ϕ_ε or, equivalently, invariant under arbitrary changes $V \to \tilde{V}$. This property we call PG independence (or CG independence if coordinates are used). According to Eq. (5.5) the condition on Q_o is $\mathcal{L}_X Q_o = 0$ for all vector fields X; this is equivalent to the following statement by Sachs (1964): The linear perturbation Q_1 of some tensorfield Q_o on a spacetime (M,g) is PG independent if and only if $Q_o = 0$ on M_o.

Examples: The curvature tensor in the linearized theory of gravity is PG independent. The perturbed curvature NP scalars Ψ_o and Ψ_4 are PG independent provided the Weyl tensor is type $\{22\}$ and a principal tetrad is used. (For a definition of principal tetrad see Sec. 5.8.)

The Linearization Procedure

A plausible way to approximate a perturbed quantity Q_ε is to obtain a solution for Q_1 by solving linearized field equations. Assume Q_ε satisfies $L(Q_\varepsilon) = 0$, where $L(Q_\varepsilon)$ is a nonlinear differential operator defined on M_ε. Now we make the expansions

$$\phi_{\varepsilon *}^{-1}\left[L(Q_\varepsilon)\right] = L_o(Q_o) + \varepsilon L(Q_o, Q_1) + O(\varepsilon^2) , \qquad (5.6)$$

where L_o is a nonlinear operator and L_1 is linear in Q_1. Inserting into $L(Q_\varepsilon) = 0$ gives

$$L_1(Q_o, Q_1) = 0 \qquad (5.7)$$

as the desired linearized perturbation equation. The splitting of $L(Q_\varepsilon)$ in the form (5.6) has to be checked in each individual case.

5.4 Point gauge dependent and point gauge independent Perturbations

A metric perturbation is never PG independent, and a connection-perturbation is not either. (In fact, even for flat space the condition $\mathcal{L}_X \eta_{ab} = 0$ would imply that all arbitrary vector fields X would have to be Killing vectors and that, in turn, implies $\eta_{ab} = 0$.) A question of basic importance for perturbation theory is therefore, to what extent can one characterize a perturbation by gauge invariant quantities? In other words, is there always a metric perturbation corresponding to a perturbation of some set of PG invariant quantities?

In view of the remarks above, one attempts to consider perturbations of gauge independent quantities only. But now one has to go backwards to find out whether there is a spacetime (i.e. a metric perturbation) that may be constructed from the PG invariant quantities; for there may well be PG invariant perturbations for which this is not the case. Let us call gauge independent perturbations for which there exists a metric <u>physical perturbations</u>; all others shall be called <u>unphysical perturbations</u>.

If we consider algebraically special spacetimes only, then there exists at least one gauge independent perturbation of one of the tetrad components of the Weyl tensor. Conversely, is there a nonempty set of unphysical perturbations? For the linearized theory of gravity the question is uniquely answered by an explicit formula given by Robinson (1972)

$$h_{ab} = 2\, x^p x^q \int_0^1 R_{apbq}(\lambda x)\, \lambda\,(1-\lambda)\, d\lambda \; + \; 2\, \xi_{(a,b)}$$

(the linearized Bianchi-indentities $R_{ab[cd;e]} = 0$ are precisely the integrability conditions guaranteeing the existence of a metric perturbation h_{ab} with $R_{abcd} = 2\varepsilon\, h_{[c,d][a,b]}$.) Whether a similar result holds for linear perturbations of curved backgrounds appears to be unknown.

Before going on, other approaches to perturbation methods are discussed.

5.5 <u>Regge-Wheeler Perturbation Technique</u>

The Regge-Wheeler perturbation method (Regge & Wheeler 1957) again considers perturbations of a metric, but replaces the Minkowski metric η by a nonflat exact solution of Einstein's vacuum equation. Thus

$$\tilde{g}_{ab} = g^{o}_{ab} + \varepsilon\, h_{ab} \, . \qquad (5.8)$$

So far the method has been applied to the case where $\overset{o}{g}_{ab}$, the metric of the unperturbed background, was either the Schwarzschild or the Reissner-Nordstrøm metric. Linearizing the field equations about $\overset{o}{g}_{ab}$ one obtains $\tilde{R}_{ab} = \overset{o}{R}_{ab} + \varepsilon R_{ab}$. Since $\overset{o}{R}_{ab}$ vanishes for a vacuum metric, R_{ab} must also vanish, which leads to a second-order differential equation for the h_{ab} (for vacuum perturbations. To obtain a solution for the h_{ab} of this equation a separation of variables ansatz may be sought in the form of a normal mode expansion. This is possible as the eigenfunctions of the equation are strongly complete (Stewart 1975) in a sense defined in Sec. 6.9. In a spherically symmetric static spacetime the coordinate functions can be adapted to the symmetries of the background geometry. Each mode of the perturbation will have an $\exp(-i\omega t)$ time dependence (in curvature coordinates t,r,θ,φ). The angular dependence separates into so-called <u>tensor spherical harmonics</u>, which form a (strongly) complete and orthonormal set of eigenfunctions on the unit sphere. They can be constructed by acting with certain first and second order differential operators on (scalar) spherical harmonics $Y_{\ell m}(\theta,\varphi)$. According to their behaviour under inversion of the radial coordinate, the normal mode components divide into two groups. Those of even (electric) parity transform as $(-1)^{\ell}$, those of odd (magnetic) parity transform as $(-1)^{\ell+1}$. After Fourier transformation of the $h_{ab}(t,r,\theta,\varphi)$ to frequency space and integration over angles one obtains a radial equation, which has to be solved in order to complete the solution.

Regge & Wheeler (1957) obtained the radial equation for the odd parity components. Further work was done by Edelstein & Vishveshwara (1970), and Vishveshwara (1970). Zerilli (1970) derived the even parity radial equation in the form of a Schrödinger equation and generalized the method to inhomogeneous perturbations, i.e. with a test particle source present. For a corrected and simplified version of Zerilli's analysis see Breuer, Tiomno & Vishveshwara (1975). For the Reissner-Nordstrøm case the perturbation equations were derived by Zerilli (1974) and Moncrief (1974a,b,c).

How one goes about analyzing a particular field in terms of vector or tensor spherical harmonics may be invariantly understood in terms of the underlying group structure. A group G of isometries (or tetrad transformations) acts on the set of fields under consideration, which form a vectorspace $\mathbb{V}$. Any element $g \in G$ commutes with the wave operator $\Box$ (or suitably chosen other operators), i.e. $g \circ \Box = \Box \circ g$. Hence any g also comutes with $\Box^{-1}$, the operation whose action one wants to know. Hence G maps the eigenspaces of $\Box^{-1}$ into itself. The eigenfunctions can be classified according to the representations of G. This gives a decomposition of $\mathbb{V}$ into a direct sum of G-invariant parts: $\mathbb{V} = \bigoplus_i V_i$. This split depends on the particular action of G and the fields under consideration.

For the Schwarzschild spacetime the isometry group is the product of time-translations and roations on the unit sphere S^2: $G = \{\text{Time-Translations}\} \times SO_3$. It is natural to study the fields using coordinates which are adapted to this group, i.e. spherical coordinates (t,r,θ,φ). The eigenfunctions of the time-translations will be $e^{-i\omega t}$ for all fields. For a scalar field, the angular eigenfunctions are scalars on S^2 under SO_3; they are the eigenfunctions of L^2, the square of the angular momentum operator, i.e. they are the $Y_{\ell m}(\theta,\varphi)$. Vector spherical harmonics may be defined in the following way. Given vectorfields $A_a(t,r,\theta,\varphi)$ and $\mathbb{V} = \{A_a(x^b)\}$, then under the action of SO_3 the components A_t, A_r transform as scalars and again the eigenfunctions are $Y_{\ell m}(\theta,\varphi)$. The quantities (A_θ, A_φ) transforms a vectorfield on S^2. One therefore takes the eigenfunctions of the part of $\Box$ which acts on this vector. In other words, vector spherical harmonics are vectorfields on S^2 which are eigenfunctions of $\Box\big|_{S^2}$.

As for tensor spherical harmonics, under SO_3 a symmetric tensor field $h_{ab}(t,r,\theta,\varphi)$, with $\mathbb{V} = \{h_{ab}(x^c)\}$ splits into scalar and vector fields as well as symmetric traceless tensor fields on S^2 [see (5.12) & (5.13) below].

In order to actually show how one decomposes a metric perturbation into tensor spherical harmonics, let us first show how a vector field $A_a(t,r,\theta,\varphi)$ decomposes into vector spherical harmonics. Its components may be obtained by the following procedure. Apply a timelike unitvector e_t, a radial unit vector e_r, the gradient $\underline{\nabla}$ and the angular momentum operator $\underline{L} = -i\,\underline{r}\times\underline{\nabla}$ to $Y_{\ell m}(\theta,\varphi)$ as follows, then

$$
\begin{aligned}
&e_t\, Y_{\ell m} = [Y_{\ell m},0,0,0]\,, \qquad e_r\, Y_{\ell m} = [0,Y_{\ell m},0,0]\,,\\
&\underline{\nabla}\, Y_{\ell m} = [0,0,\partial_\theta Y_{\ell m}, \partial_\varphi Y_{\ell m}]\,, \qquad (5.9)\\
&\underline{L}\, Y_{\ell m} = [0,0,-\frac{1}{\sin\theta}\partial_\varphi Y_{\ell m}, \sin\theta\, \partial_\theta Y_{\ell m}]\,.
\end{aligned}
$$

Whereas e_t, e_r and $\underline{\nabla}$ produce even parity components ($Y_{\ell m}$ by it-

self is of even parity), $\underline{L}$ yields an odd parity component. Hence a complete normal mode expansion of $A_a = A_a^{odd} + A_a^{even}$ is

$$A_a(t,r,\theta,\varphi) = \sum_{\ell,m} \left\{ \begin{pmatrix} 0 \\ 0 \\ -\frac{a_o}{\sin\theta}\partial_\varphi Y_{\ell m} \\ a_o \sin\theta\, \partial_\theta Y_{\ell m} \end{pmatrix} + \begin{pmatrix} a_1 Y_{\ell m} \\ a_2 Y_{\ell m} \\ a_3 \partial_\theta Y_{\ell m} \\ a_3 \partial_\varphi Y_{\ell m} \end{pmatrix} \right\}, \tag{5.10}$$

where the $a_i \equiv a_i(t,r)$ are arbitrary functions. Considering (5.10) as the electromagnetic potential and expanding the four-current j_a in exactly the same manner in a spherically symmetric background, Maxwell's equations separate (Ruffini, Tiomno & Vishveshwara 1972). They give rise to second differential equations for $a_o(t,r)$ as well as for an even parity radial function, which represent the two transverse degrees of freeedom of the electromagnetic field. In the homogeneous case they both satisfy the same equation, first found by Wheeler (1962), namely

$$-\frac{d^2R_{\ell m}(r)}{dr^{*2}} + \left[(1-\frac{2M}{r})\frac{\ell(\ell+1)}{r^2} - \omega^2\right]R_{\ell m}(r) = 0. \tag{5.11}$$

In order to expand a tensor field $h_{ab}(t,r,\theta,\varphi) = h_{ba}$ into tensor spherical harmonics one constructs second order operators like $\underline{L}$ out of the ones given in (5.9). For the most general odd parity perturbation the Regge-Wheeler form is

$$h_{ta}^{odd} = h_o(t,r)\left[0,0,-\frac{1}{\sin\theta}\partial_\varphi Y_{\ell m}, \sin\theta\,\partial_\theta Y_{\ell m}\right] \tag{5.12a}$$

and

$$h_{ij}^{odd} = h_1(t,r)(\hat{e}_1)_{ij} + h_2(t,r)(\hat{e}_2)_{ij}, \quad (i,j = r,\theta,\varphi) \tag{5.12b}$$

where

$$\hat{e}_1 = \begin{bmatrix} 0 & -\frac{1}{\sin\theta}\partial_\varphi Y_{\ell m} & \sin\theta\,\partial_\theta Y_{\ell m} \\ * & 0 & 0 \\ * & 0 & 0 \end{bmatrix}$$

and

$$\hat{e}_2 = \begin{bmatrix} 0 & 0 & 0 \\ 0 & \frac{1}{\sin\theta}\left[\partial_\theta\partial_\varphi - \cot\theta\,\partial_\varphi\right]Y_{\ell m} & * \\ 0 & \frac{1}{2}\left[\frac{1}{\sin\theta}\partial_\varphi^2 + \cos\theta\,\partial_\theta - \sin\theta\,\partial_\theta^2\right]Y_{\ell m} & -\left[\sin\theta\,\partial_\theta\partial_\varphi - \cos\theta\,\partial_\varphi\right]Y_{\ell m} \end{bmatrix}$$

and $*$ indicates a symmetric component. For terms of even parity Regge & Wheeler write the most general decomposition as

$$h_{ta}^{even} = \left[(1 - \frac{2M}{r})H_o(t,r)Y_{\ell m}, H_1(t,r)Y_{\ell m}, h_o(t,r)\,\partial_\theta Y_{\ell m}, h_o(t,r)\partial_\varphi Y_{\ell m}\right], \tag{5.13a}$$

and

$$h_{ij}^{even} = h_1(t,r)(\hat{f}_1)_{ij} + \frac{H_2(t,r)}{1-2M/r}(\hat{f}_2)_{ij} + r^2K(t,r)(\hat{f}_3)_{ij} + r^2G(t,r)(\hat{f}_4)_{ij}, \tag{5.13b}$$

where

$$\hat{f}_1 = \begin{bmatrix} 0 & Y_{\ell m} & Y_{\ell m} \\ * & 0 & 0 \\ * & 0 & 0 \end{bmatrix}, \hat{f}_2 = \begin{bmatrix} Y_{\ell m} & 0 & 0 \\ 0 & 0 & 0 \\ 0 & 0 & 0 \end{bmatrix}, \hat{f}_3 = \begin{bmatrix} 0 & 0 & 0 \\ 0 & Y_{\ell m} & 0 \\ 0 & 0 & \sin\theta\, Y_{\ell m} \end{bmatrix},$$

$$\hat{f}_4 = \begin{bmatrix} 0 & 0 & 0 \\ 0 & \partial_\theta^2 Y_{\ell m} & * \\ 0 & \left[\partial_\theta\partial_\varphi - \cot\theta\,\partial_\varphi\right]Y_{\ell m} & \left[\partial_\varphi^2 + \sin\theta\ \cos\theta\,\partial_\theta\right]Y_{\ell m} \end{bmatrix}.$$

Furthermore, by a suitable choice of gauge, e.g. the "transverse traceless gauge", the number of elements of the perturbation matrix can be reduced and one arrives at second order differential equations for the two independent radial functions $R_{\ell m}^{even}(r)$ and $R_{\ell m}^{odd}(r)$ for even and odd parities, respectively. In the sourceless case, they satisfy the equations

$$-\frac{d^2R_{\ell m}(r)}{dr^{*2}} + \left[V_\ell(r) - \omega^2\right]R_{\ell m}(r) = 0\ , \tag{5.14}$$

where

$$V_\ell^{even}(r) = (1 - \frac{2M}{r})\ \frac{2\lambda^2(\lambda+1)r^3 + 6\lambda Mr^2 + 18\lambda M^2r + 18M^3}{r^3(\lambda r + 3M)^2}\ ,$$

$$V_{\ell}^{odd}(r) = (1 - \frac{2M}{r})\left[\frac{\ell(\ell+1)}{r^2} - \frac{6M}{r^3}\right],$$

$$\lambda = \frac{1}{2}(\ell-1)(\ell+2)\ , \quad dr/dr^* = 1 - 2M/r.$$

For a test gravitational field produced by a test particle the source term T_{ab} is given by an integral over the world line of the particle, the particle being represented by a four-dimensional δ-function The source term is divergence-free if the world line is a geodesic of the background geometry. The inhomogeneous perturbation equations are obtained by linearizing Einstein's equations $\tilde{G}(\tilde{g}_{ab}) = 8\pi\,\tilde{T}_{ab}$. One proceeds as outlined in Eqs. (5.6 & 7); linearizing the metric $\tilde{g}$ and the operator $L \equiv \tilde{G}$ one obtains

$$\tilde{g} = g^o + \varepsilon\, h, \quad \tilde{G} = \mathcal{G}_o + \varepsilon\, \mathcal{G}_1,$$

$$\tilde{G}(\tilde{g}) = \phi_{\varepsilon *}\left[G_o(g^o) + \varepsilon\, G_1(g^o,h)\right]$$

$$= 0 + \varepsilon\, \phi_{\varepsilon *}\left[G_1(g^o)[h]\right], \tag{5.15a}$$

$$\tilde{T} = T^o + \varepsilon\, T_1 = 0 + \varepsilon\, T_1\ . \tag{5.15b}$$

Hence

$$G_1(g^o)\,[h] = 8\pi\, T_1 \tag{5.15c}$$

is the desired equation, where $G_1(g)$ is a linear differential operator acting on h, and T_1 is given by the delta-function source which has to be expanded in tensor spherical harmonics in a similar manner.

Up to date this machinery has been applied to the Schwarzschild and Reissner-Nordstrøm geometries as background only. The corresponding scalar (see, e.g., Matzner 1968, Breuer et al. 1973c) and electromagnetic wave equations (Ruffini et al. 1972) have been separated by this method. Also the scalar wave equation for a Kerr background (Brill et al. 1972) has been separated with the aid of spheroidal angular harmonics. The electromagnetic and gravitational wave equations for the Kerr geometry as background have not yet been dealt with in this approach, due to the complexities in constructing and handling vector and tensor spheroidal harmonics.

The Regge-Wheeler approach naturally suffers from the same drawbacks as the linearized theory, since analogous problems of gauge dependence arise. In addition the formalism has to be worked out separately for each different metric and spin and presence of a certain symmetry group. It is interesting to note, however, that despite the gauge dependence of the perturbations, one is able to choose suitable combinations of coordinate functions and their derivatives which are gauge independent and which also lead to separated equations (cf. Moncrief 1974 a,b,c for Reiss-Nordstrøm metric, and Chrzanowski 1975, Chandrasekhar 1975 for the Schwarzschild metric.).

A different perturbation approach which considers solely gauge independent perturbations can be set up using the formalism of Newman & Penrose (see Sec. 5.8). But first a brief discussion of the Regge-calculus will be given.

5.7 Perturbation Theory and Regge-Calculus

As a standard tool in numerical analysis one replaces partial differential equations by difference equations, i.e. one replaces continuous space by a grid of points and derivatives by differences. This idea has been applied to the theory of spacetimes some time ago (Silberstein 1936; Schild 1948). Demanding invariance of such a 4-dimensional lattice under Lorentz transformations one always obtains - due to the discreteness - a minimal discrete Lorentz transformation associated with a minimal non-zero velocity of $\frac{3}{2}c = 0.866\ c$. This drawback is lethal for this type of approximation method.

A more successful kind of discretization was introduced by Regge (1961) (see MTW, Chap. 42). In his approach spacetime is approximated by a net of simplicial cells, which in two, three or four dimensions are triangles, tetrahedrons or four-simplices, respectively. These cells are joined at the vertices called "bones" and the resulting geometry is called a "skeleton geometry". Within the skeleton geometry one looks for analogs of the quantities and operations that occur in the continuous geometry. The cells are intrinsically flat, except for the bones which represent curvature. The deviation from flatness at the vertex can be measured by the "deficit angle δ" of a polyhedron if this vertex were laid out on a flat hypersurface, as shown in the following picture.

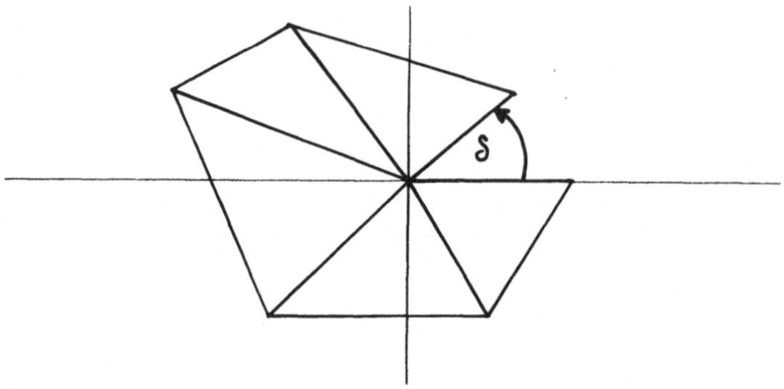

Einstein's equations may now be reexpressed in terms of constraints on the lengths of the edges of the cells. Advantages of this technique are that it allows coordinate and gauge free formulations of the problems under consideration. Also, the symmetries of spacetimes play no essential role in the formalism; hence more complicated geometries can equally well be treated. In engineering a quite similar approach has been developed under the name of the "finite element method" (Desai & Abel 1972).

This technique has been applied to finding solutions of the initial value problem in the time symmetric case (Wong 1971, Collins & Williams 1972) and to simple dynamical situations such as the Friedmann- and the Tolman universe (Collins & Williams 1973, 1974). In the last two cases the time dependence, i.e. expansions or contractions of the universe, are expressed in terms of the variation of the edge-length with time. Recently, using Regge calculus, Sorkin (1974) has formulated the general time evolution and initial-value problems. Existence and stability of spacetime skeletons have, however, not yet been studied.

The reason for mentioning Regge calculus here is the hope that radiation theory may equally be treated as a perturbation of skeleton spacetimes. The propagation equations for such infinitesimal perturbations would be formulated as refraction laws at the bones. Through the interior of the cells radiation propagates as in flat spacetime.

It is not known whether such an approach is tractable and effective. The inherent gauge independence of the formalism, however, at least provides a natural starting point. If the approach works it can be applied to a wider range of geometries than so far dealt with by conventional perturbation theory.

5.8 The Newman-Penrose (NP) Formalism

The formalism of Newman & Penrose (1962) is a null tetrad calculus. It is related in a natural way to the spinor formulation of General Relativity. One introduces a null tetrad $\{\ell, n, m, \bar{m}\}$ which consists of two real null vectors ℓ, n and one complex spacelike vector m, together with its complex conjugate $\bar{m}$ satisfying the orthonormality relations

$$-\ell \cdot n = +m \cdot \bar{m} = 1, \quad \text{all other products} = 0, \tag{5.16}$$

$$\Leftrightarrow \qquad g_{ab} = -2\left[\ell_{(a}n_{b)} - m_{(a}\bar{m}_{b)}\right] \quad .$$

The vector m can be constructed from two real, orthogonal, spacelike unit vectors e_1 and e_2 as $m = 2^{-1/2}(e_1 - ie_2)$. Four directional derivatives along the tetrad legs are defined by

$$D \equiv \ell^a \nabla_a \quad , \qquad \Delta \equiv n^a \nabla_a \quad ,$$

$$\delta \equiv m^a \nabla_a \quad , \qquad \bar{\delta} \equiv \bar{m}^a \nabla_a \quad . \tag{5.17}$$

Twelve complex functions, the <u>spin coefficients</u> (which are in fact the tetrad components of the Ricci rotation coefficients) are then defined by

$$\alpha = \tfrac{1}{2}(n^a \bar{m}^b \nabla_b \ell_a - \bar{m}^a \bar{m}^b \nabla_b m_a) \, ,$$

$$\beta = \tfrac{1}{2}(n^a m^b \nabla_b \ell_a - \bar{m}^a m^b \nabla_b m_a) \, ,$$

$$\gamma = \tfrac{1}{2}(n^a n^b \nabla_b \ell_a - \bar{m}^a n^b \nabla_b m_a) \, ,$$

$$\varepsilon = \tfrac{1}{2}(n^a \ell^b \nabla_b \ell_a - m^a \ell^b \nabla_b m_a) \, ,$$

$$\lambda = -\bar{m}^a \bar{m}^b \nabla_b n_a \, , \qquad \mu = -\bar{m}^a m^b \nabla_b n_a \, , \qquad \nu = -\bar{m}^a n^b \nabla_b n_a \, ,$$

$$\pi = -\bar{m}^a \ell^b \nabla_b n_a \, , \qquad \kappa = m^a \ell^b \nabla_b \ell_a \, , \qquad \rho = m^a \bar{m}^b \nabla_b \ell_a \, ,$$

$$\sigma = m^a m^b \nabla_b \ell_a \, , \qquad \tau = m^a n^b \nabla_b \ell_a \, .$$

Again, these definitions are given more naturally in terms of spinors. These spin coefficients, then, are simply the components of the directional derivatives of the basis spinors that are equivalent to the tetrad $\{\ell, n, m, \bar{m}\}$. The ten independent tetrad components of the Weyl tensor C_{abcd} can be written as five complex scalars, namely

$$\Psi_0 = - C_{abcd}\, \ell^a m^b \ell^c m^d , \qquad \Psi_1 = - C_{abcd}\, \ell^a n^b \ell^c m^d ,$$

$$\Psi_2 = - \tfrac{1}{2} C_{abcd}\, (\ell^a n^b \ell^c n^d + \ell^a n^b m^c \bar{m}^d) , \tag{5.19}$$

$$\Psi_3 = - C_{abcd}\, \ell^a n^b \bar{m}^c n^d , \qquad \Psi_4 = - C_{abcd}\, n^a \bar{m}^b n^c \bar{m}^d .$$

Writing the Ricci identities and the Bianchi identities in this formalism one obtains a set of equations, loosely called the "field equattions", which in this formalism take the place of Einstein's equations.

For an electromagnetic field the six independent tetrad components of F_{ab} can be replaced by three complex scalars,

$$\varphi_0 = F_{ab}\, \ell^a m^b , \quad \varphi_1 = \tfrac{1}{2} F_{ab} (\ell^a n^b + m^a \bar{m}^b) , \quad \varphi_2 = F_{ab}\, \bar{m}^a n^b . \tag{5.20}$$

The tetrad components of Maxwell's equations become a set of four equations for these scalars.

For references see Newman & Penrose (1962), Newman & Unti (1962), and especially the review article by Pirani (1964). Perturbation equations can now be obtained by writing all NP quantities and operators as, for example, $\tilde{\Psi} = \Psi^0 + \varepsilon\Psi$, $\tilde{D} = D^0 + \varepsilon D$, $\tilde{\kappa} = \kappa^0 + \varepsilon\kappa$ etc. and expanding the full equations to first order in ε .

5.9 General Relativity and the Formalism of Geroch, Held and Penrose (GHP)

In the NP formalism one has to choose a complete null tetrad at each point of spacetime. In general, this choice is somewhat arbitrary when the geometry does not suggest a natural choice for the tetrad vectors. In all algebraically special spacetimes, however, two null directions can be distinguished at each point, one of these being the repeated principal null direction of the Weyl tensor. The second null direction can be specified essentially by the condition that the distribution of timelike two-surfaces defined by the null directions be integrable (Walker & Held 1974). The GHP formalism (Geroch, Held & Penrose 1973) is tailored to this situation and constitutes a simplification of the NP method.

In a type $\{22\}$ spacetime one can take ℓ and n of the null tetrad to be the two repeated null directions of the Weyl tensor. Such a tetrad shall be called a <u>principal</u> <u>tetrad</u>. In fact, all principal tetrads are the same modulo a boost as given in (5.21). The orthonormality conditions (5.16) are unchanged under transformations which leave the directions of ℓ and n invariant. These are rotations in the $m - \bar{m}$ plane through an angle θ and boosts of magnitude r in the $\ell - n$ plane,

$$\ell \rightarrow r\ell \ , \quad n \rightarrow r^{-1}n, \quad m \rightarrow e^{i\theta}m \ . \tag{5.21}$$

The essential idea is to consider only quantities Q which transform homogeneously under these transformations. A quantity Q is said to be <u>of type (p,q)</u> or a (p,q) quantity if under transformation (5.21) it transforms according to

$$Q \rightarrow r^{p+q}\ e^{i(p-q)}\ Q \equiv \lambda^p\ \bar{\lambda}^q\ Q \quad ,$$

where $\lambda = re^{i\theta}$. Similarly, as in the Regge-Wheeler formalism, one now considers the vectorspace $\mathbb{V}$ of all quantities Q, $\mathbb{V} = \{Q\}$ acted upon by the multiplicative group of complex numbers $G = \{\lambda, \lambda \in \mathbb{C}\}$. According to the above action, one finds the irreducible parts $\mathbb{V}(p,q)$ of $\mathbb{V} = \bigoplus_{p,q} \mathbb{V}(p,q)$ each of which contains all quantities of type (p,q). In terms of <u>spin-weight</u> <u>s</u> and <u>boost-weight</u> <u>w</u> a type is then expressed as

$$s = p - q \ , \qquad w = p + q$$

This notation is more transparent when the tetrad vectors are expressed in terms of a spinor basis o^A and ι^A, namely by $\ell^a = o^A o^{A'}$, $n^a = \iota^A \iota^{A'}$, $m^a = o^A \iota^{A'}$. Under transformations leaving the directions of and n invariant this basis is unique up to changes $o^A \rightarrow \lambda o^A$ and $\iota^A \rightarrow \lambda^{-1}\iota^A$. This leads immediately to $\ell^a \rightarrow \lambda\bar{\lambda}\,\ell^a$ and hence ℓ is of type $(1,1)$ etc.

Some quantities transform inhomogeneously under (5.21) and for them a type cannot be defined. The only NP scalars for which this is the case are $\alpha, \beta, \gamma, \varepsilon$ in (5.18). For the operators D, Δ and δ [see Eq. (5.17)] a type cannot, in general be defined either. I.e., these operators do not commute with the group action. For example, if Q has a certain type, then it is in general impossible to define a type for DQ. However, one can introduce a new operator

$$\Theta_a = \nabla_a - \tfrac{1}{2}(p+q)\,\alpha_a + \tfrac{1}{2}(p-q)\,\beta_a\,, \qquad (5.22)$$

with $\alpha_a = n^c \nabla_a \ell_c$ and $\beta_a = \bar{m}^c \nabla_a m_c$. This operator is of type (0,0). Transvecting it with the tetrad vectors leads to the following new operators:

$$\text{þ} = D - p\varepsilon - q\bar{\varepsilon}\,, \quad \text{þ}' = \Delta - p\gamma - q\bar{\gamma}\,, \quad \text{(pronounced "thorn")}$$
$$\eth = \delta - p\beta - q\bar{\alpha}\,, \quad \eth' = \bar{\delta} - p\alpha - q\bar{\beta}\,, \quad \text{(pronounced "edth")} \qquad (5.23)$$

for which a type may always be defined. They act on quantities of type (p,q). Only when applied to a (0,0) quantity do the GHP operators become identical with the NP operators.

A further simplification in the GHP formalism occurs when one introduces the following three symmetry operations, under which the full set of NP equations is invariant:

	complex conjugation $^-$	prime $'$	star $*$
ℓ	ℓ	n	m
n	n	ℓ	$-\bar{m}$
m	$\bar{m}$	$\bar{m}$	$-\ell$
$\bar{m}$	m	m	n

If a quantity has type (p,q), then these operations change it, respectively, to

$$^- : (p,q) \rightarrow (q,p)\,, \quad ' : (p,q) \rightarrow (-p,-q)\,, \quad * : (p,q) \rightarrow (p,-q)\,. \qquad (5.24)$$

Complex conjugation and the prime operation commute; applying a prime twice gives the identity. We have

$$(\bar{\Omega})' = \overline{(\Omega')}, \quad \Omega'' = \Omega, \quad \Omega^{**} = (-1)^q \, \Omega ,$$
$$(\Omega')^* = (-1)^q \, (\Omega^*)', \quad \overline{(\Omega^*)} = (-i)^{p+q} (\bar{\Omega}')^* . \tag{5.25}$$

If one is given a quantity Ω, then by applying all combinations of symmetry operations we can in general derive six other quantities, namely $\bar{\Omega}$, Ω', Ω^*, $(\bar{\Omega})^*$, $(\Omega')^*$ and $(\bar{\Omega}')^*$. The effect of these operations on the spin coefficients and the operators is exhibited in Table II. The type for these various quantities can also be read off Fig. 5.

The NP field scalars φ_0, φ_1, φ_2 and $\Psi_0, \ldots, \Psi_4$ as defined in (5.19 & 20) are of type $(2s-2n, 0)$, where $s = 1$, $n = 0,1,2$ for the φ_n and $s = 2$, $n = 0,\ldots,4$ for the Ψ_n. The scalar field corresponds to the case $s = n = 0$. The tetrad components of any sources are defined by

$$J_\ell = J_a \ell^a, \quad J_n = J_a n^a, \text{ etc. for Electromagnetism}$$

and

$$T_{\ell\ell} = T_{ab} \ell^a \ell^b, \quad T_{\ell n} = T_{ab} \ell^a n^b, \quad \text{etc. for GR.}$$

The corresponding tetrad components of the Ricci tensor are $-\frac{1}{2} R_{\ell\ell} = \Phi_{00} = 4\pi T_{\ell\ell}$, etc. (see Pirani 1964; for the types see again Table II, where Einstein's equations have been used.) Now the NP equations can be written down in the GHP language.

<u>Maxwell Equations</u>

In this notation the Maxwell equations with source reduce to one equation plus three others that can be derived from it by GHP symmetry operations using Table II:

$$þ\varphi_1 - ð'\varphi_0 = -\tau'\varphi_0 + 2\rho\varphi_1 - \kappa\varphi_2 + 2\pi J_\ell \, , \tag{5.26}$$
$$þ'\varphi_1 - ð\varphi_2 = -\tau\varphi_2 + 2\rho'\varphi_1 - \kappa'\varphi_0 - 2\pi J_n \, , \qquad -(5.26)'$$
$$ð\varphi_1 - þ'\varphi_0 = -\rho'\varphi_0 + 2\tau\varphi_1 - \sigma\varphi_2 + 2\pi J_m \, , \qquad (5.26)^*$$
$$ð'\varphi_1 - þ\varphi_2 = -\rho\varphi_2 + 2\tau'\varphi_1 - \sigma'\varphi_0 - 2\pi J_{\bar{m}} \, . \qquad -(5.26)^{*\prime}$$

Table II. Effect of prime, star and complex conjugation on the GHP quantities

Q	Type	Q'	Q^*	$\bar{Q}^*$	Q'^*	$\bar{Q}'^*$
ρ	(1,1)	$\rho' = -\mu$	τ	$\bar{\tau}'$	$-\tau'$	$-\bar{\tau}$
τ	(1,-1)	$\tau' = -\pi$	$-\rho$	$\bar{\rho}'$	ρ'	$-\bar{\rho}$
κ	(3,1)	$\kappa' = -\nu$	σ	$-\bar{\sigma}'$	$-\sigma'$	$\bar{\sigma}$
σ	(3,-1)	$\sigma' = -\lambda$	$-\kappa$	$-\bar{\kappa}'$	κ'	$-\bar{\kappa}$
ε	-	$\varepsilon' = -\gamma$	β	$\bar{\beta}'$	$-\beta'$	$-\bar{\beta}$
β	-	$\beta' = -\alpha$	ε	$\bar{\varepsilon}'$	ε'	$-\bar{\varepsilon}$
þ	(1,1)	$\text{þ}'$	ð	ð	$-\text{ð}'$	$-\text{ð}'$
ð	(1,-1)	$\text{ð}'$	$-\text{þ}$	þ	$\text{þ}'$	$-\text{þ}$
l	(1,1)	n	m	$-m$	$-\bar{m}$	$-\bar{m}$
n	(-1,-1)	l	$-\bar{m}$	$-\bar{m}$	m	m
m	(1,-1)	$\bar{m}$	$-l$	n	n	$-l$
$\bar{m}$	(-1,1)	m	n	l	$-l$	n
Φ	(0,0)	Φ	Φ	$\bar{\Phi}$	Φ	$\bar{\Phi}$
φ_0	(2,0)	$-\varphi_2$	φ_0	$-\bar{\varphi}_2$	$-\varphi_2$	$-\bar{\varphi}_0$.
φ_1	(0,0)	$-\varphi_1$	φ_1	$\bar{\varphi}_1$	$-\varphi_1$	$-\bar{\varphi}_1$
φ_2	(-2,0)	$-\varphi_0$	φ_2	$-\bar{\varphi}_0$	$-\varphi_0$	$-\bar{\varphi}_2$
Ψ_0	(4,0)	Ψ_4	Ψ_0	$\bar{\Psi}_4$	Ψ_4	$\bar{\Psi}_0$
Ψ_1	(2,0)	Ψ_3	Ψ_1	$-\bar{\Psi}_3$	Ψ_3	$-\bar{\Psi}_1$
Ψ_2	(0,0)	Ψ_2	Ψ_2	$\bar{\Psi}_2$	Ψ_2	$\bar{\Psi}_2$
Ψ_3	(-2,0)	Ψ_1	Ψ_3	$-\bar{\Psi}_1$	Ψ_1	$-\bar{\Psi}_3$
Ψ_4	(-4,0)	Ψ_0	Ψ_4	$\bar{\Psi}_0$	Ψ_0	$\bar{\Psi}_4$

Table II (cont'd). Effect of prime, star and complex conjugation on the sources

Q	Type	Q'	Q^*	$\bar{Q}^*$	Q'^*	$\bar{Q}'^*$	Ricci scalar/4π
J_ℓ	(1,1)	J_n	J_m	J_m	$-J_{\bar{m}}$	$-J_{\bar{m}}$	-
J_n	(-1,-1)	$-J_\ell$	$-J_{\bar{m}}$	$-J_{\bar{m}}$	J_m	J_m	-
J_m	(1,-1)	$J_{\bar{m}}$	$-J_\ell$	J_n	J_n	$-J_\ell$	-
$J_{\bar{m}}$	(-1,1)	J_m	J_n	J_ℓ	$-J_\ell$	J_n	-
$T_{\ell\ell}$	(2,2)	T_{nn}	T_{mm}	T_{mm}	$T_{\bar{m}\bar{m}}$	$T_{\bar{m}\bar{m}}$	Φ_{00}
$T_{\ell m}$	(2,0)	$T_{n\bar{m}}$	$-T_{\ell m}$	$-T_{n\bar{m}}$	$-T_{n\bar{m}}$	$-T_{\ell\bar{m}}$	Φ_{01}
$T_{\ell\bar{m}}$	(0,2)	T_{nm}	T_{nm}	$T_{\ell m}$	$T_{\ell\bar{m}}$	$-T_{n\bar{m}}$	Φ_{10}
$T_{\ell n}$	(0,0)	$T_{\ell n}$	$-T_{m\bar{m}}$	$T_{m\bar{m}}$	$-T_{m\bar{m}}$	$-T_{m\bar{m}}$	$\}\ \frac{1}{2}(T_{\ell n}+T_{m\bar{m}}) = \Phi_{11}$
$T_{m\bar{m}}$	(0,0)	$T_{m\bar{m}}$	$-T_{\ell n}$	$-T_{\ell n}$	$-T_{\ell n}$	$-T_{\ell n}$	
T_{nm}	(0,-2)	$T_{\ell\bar{m}}$	$T_{\ell\bar{m}}$	$-T_{n\bar{m}}$	T_{nm}	$-T_{\ell m}$	Φ_{12}
$T_{n\bar{m}}$	(-2,0)	$T_{\ell m}$	$-T_{n\bar{m}}$	$-T_{\ell\bar{m}}$	$-T_{\ell m}$	T_{nm}	Φ_{21}
T_{nn}	(-2,-2)	$T_{\ell\ell}$	$T_{\bar{m}\bar{m}}$	$-T_{\bar{m}\bar{m}}$	T_{mm}	T_{mm}	Φ_{22}
T_{mm}	(2,-2)	$T_{\bar{m}\bar{m}}$	$T_{\ell\ell}$	T_{nn}	T_{nn}	$T_{\ell\ell}$	Φ_{02}
$T_{\bar{m}\bar{m}}$	(-2,2)	T_{mm}	T_{nn}	$T_{\ell\ell}$	$T_{\ell\ell}$	T_{nn}	Φ_{20}
$\Lambda = R/24$	(0,0)	Λ	Λ	Λ	Λ	Λ	-

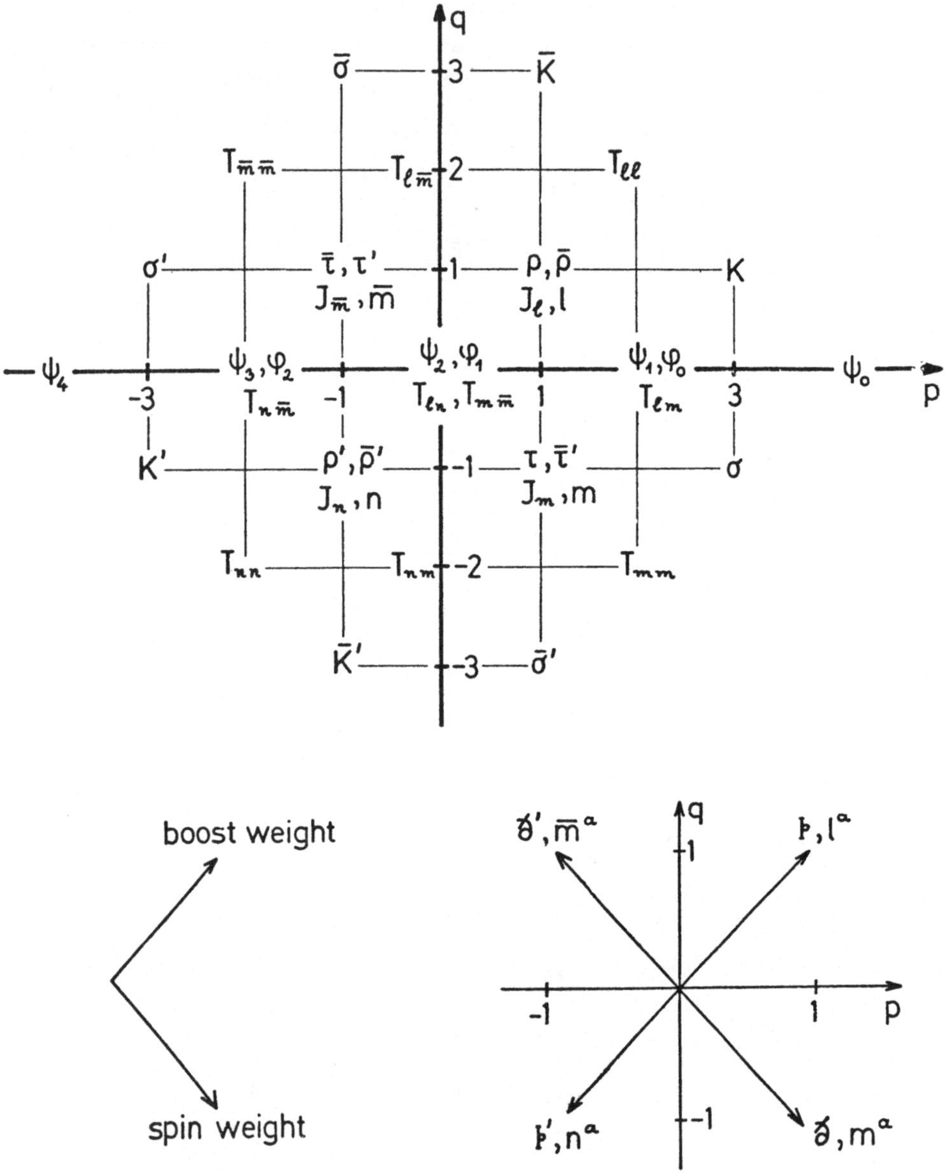

Figure 6. Types of GHP quantities

Field Equations

Similarly, the inhomogeneous field equations consist of a set of four equations plus eight others derivable from them:

$$\eth\rho - \eth'\sigma = (\rho - \bar\rho)\tau + (\bar\rho' - \rho')\kappa - \Psi_1 + 4\pi T_{lm} , \qquad (5.27)$$

$$\eth'\rho' - \eth\sigma' = (\rho' - \bar\rho')\tau' + (\bar\rho - \rho)\kappa' - \Psi_3 + 4\pi T_{n\bar m} , \qquad (5.27)'$$

$$\text{þ}\tau - \text{þ}'\kappa = (\tau - \bar\tau')\rho + (\tau - \tau')\sigma + \Psi_1 + 4\pi T_{lm} , \qquad -(5.27)^*$$

$$\text{þ}'\tau' - \text{þ}\kappa' = (\tau' - \bar\tau)\bar\rho + (\bar\tau' - \tau)\sigma' + \Psi_3 + 4\pi T_{n\bar m} ; \qquad -(5.27)^{*\prime}$$

$$\text{þ}\rho - \eth'\kappa = \rho^2 + \sigma\bar\sigma - \bar\kappa\tau - \kappa\tau' + 4\pi T_{ll} , \qquad (5.28)$$

$$\text{þ}'\rho' - \eth\kappa' = \rho'^2 + \sigma'\bar\sigma' - \bar\kappa'\tau' - \kappa'\tau + 4\pi T_{nn} , \qquad (5.28)'$$

$$\eth\tau - \text{þ}'\sigma = \tau^2 + \kappa\bar\kappa' - \bar\sigma'\rho - \rho'\sigma + 4\pi T_{mm} , \qquad (5.28)^*$$

$$\eth'\tau' - \text{þ}\sigma' = \tau'^2 + \kappa'\bar\kappa - \bar\sigma\rho' - \rho\sigma' + 4\pi T_{\bar m\bar m} ; \qquad (5.28)^{*\prime}$$

$$\text{þ}\sigma - \eth\kappa = (\rho + \bar\rho)\sigma - (\tau + \bar\tau')\kappa + \Psi_0 , \qquad (5.29)$$

$$\text{þ}'\sigma' - \eth'\kappa' = (\rho' + \bar\rho')\sigma' - (\tau' + \bar\tau)\kappa' + \Psi_4 ; \qquad (5.29)'$$

$$\text{þ}'\rho - \eth'\tau = \rho\bar\rho' + \sigma\sigma' - \tau\bar\tau - \kappa\kappa' - \Psi_2 - 2\Lambda , \qquad (5.30)$$

$$\text{þ}\rho' - \eth\tau' = \rho'\bar\rho + \sigma'\sigma - \tau'\bar\tau' - \kappa'\kappa - \Psi_2 - 2\Lambda . \qquad (5.30)'$$

The quantity Λ is defined in Table II. Applying other GHP symmetry operations to Eqs. (5.26 - 30) yields nothing new.

Bianchi Identities

The Bianchi identities, given here for the most general case, can be derived from two equations by symmetry operations:

$$(\eth' - \tau')\Psi_0 - (\text{þ} - 4\rho)\Psi_1 - 3\kappa\Psi_2 = 4\pi[(\eth - \bar{\tau}')T_{ll} - (\text{þ} - 2\bar{\rho})T_{lm} + 2\sigma T_{l\bar{m}} - \kappa(T_{ln} + T_{m\bar{m}}) - \bar{\kappa}T_{mm}] , \tag{5.31}$$

$$(\eth - \tau)\Psi_4 - (\text{þ}' - 4\rho')\Psi_3 - 3\kappa'\Psi_2 = 4\pi[(\eth' - \bar{\tau})T_{nn} - (\text{þ}' - 2\bar{\rho}')T_{n\bar{m}} + 2\sigma' T_{nm} - \kappa'(T_{ln} + T_{m\bar{m}}) - \bar{\kappa}' T_{\bar{m}\bar{m}}] , \tag{5.31$'$}$$

$$(\text{þ}' - \rho')\Psi_0 - (\eth - 4\tau)\Psi_1 - 3\sigma\Psi_2 = -4\pi[(\text{þ} - \bar{\rho})T_{mm} - (\eth' - 2\bar{\tau})T_{n\bar{m}} + 2\kappa T_{nm} - \sigma(T_{ln} + T_{m\bar{m}}) - \bar{\sigma}' T_{ll}] , \tag{5.31*}$$

$$(\text{þ} - \rho)\Psi_4 - (\eth' - 4\tau')\Psi_3 - 3\sigma'\Psi_2 = -4\pi[(\text{þ}' - \bar{\rho}')T_{\bar{m}\bar{m}} - (\eth - 2\bar{\tau}')T_{lm} + 2\kappa' T_{l\bar{m}} - \sigma'(T_{ln} + T_{m\bar{m}}) - \bar{\sigma}T_{nn}] , \tag{5.31$^{*\prime}$}$$

$$\sigma'\Psi_0 + (\eth' - 2\tau')\Psi_1 - (\text{þ} - 3\rho)\Psi_2 - 2\kappa\Psi_3 = \frac{4\pi}{3}[(\eth' - 2\bar{\tau} + 2\tau')T_{lm} - (\text{þ}' - \bar{\rho}' + 2\rho')T_{ll} - 2(\eth - \bar{\tau}' + \tau)T_{l\bar{m}} + (\text{þ} - 2\bar{\rho} + \rho)(T_{ln} + T_{m\bar{m}}) - 2\sigma T_{\bar{m}\bar{m}} + \bar{\sigma}T_{mm} + 2\bar{\kappa}T_{nm} + 2\kappa T_{n\bar{m}}] , \tag{5.32}$$

$$\sigma\Psi_4 + (\eth - 2\tau)\Psi_3 - (\text{þ}' - 3\rho')\Psi_2 - 2\kappa'\Psi_1 = \frac{4\pi}{3}[(\eth - 2\bar{\tau}' + 2\tau)T_{n\bar{m}} - (\text{þ} - \bar{\rho} + 2\rho)T_{nn} - 2(\eth' - \bar{\tau} + \tau')T_{n\bar{m}} + (\text{þ}' - 2\bar{\rho}' + \rho')(T_{ln} + T_{m\bar{m}}) - 2\sigma' T_{mm} + \bar{\sigma}' T_{\bar{m}\bar{m}} + 2\bar{\kappa}' T_{l\bar{m}} + 2\kappa' T_{lm}] , \tag{5.32$'$}$$

$$\kappa'\Psi_0 + (\text{þ}' - 2\rho')\Psi_1 - (\eth - 3\tau)\Psi_2 - 2\sigma\Psi_3 = -\frac{4\pi}{3}[(\text{þ} - 2\bar{\rho} + 2\rho)T_{n\bar{m}} - (\eth' - \tau + 2\tau')T_{nn} - 2(\text{þ} - \bar{\rho} + \rho)T_{nm} + (\eth - 2\bar{\tau}' + \tau)(T_{ln} + T_{m\bar{m}}) - 2\kappa T_{nn} + \bar{\kappa}' T_{ll} + 2\bar{\sigma}' T_{l\bar{m}} + 2\sigma T_{n\bar{m}}] , \tag{5.32*}$$

$$\kappa\Psi_4 + (\text{þ} - 2\rho)\Psi_3 - (\eth' - 3\tau')\Psi_2 - 2\sigma'\Psi_1 = -\frac{4\pi}{3}[(\text{þ}' - 2\bar{\rho}' + 2\rho')T_{lm} - (\eth - \tau' + 2\tau)T_{ll} - 2(\text{þ}' - \bar{\rho}' + \rho')T_{l\bar{m}} + (\eth' - 2\bar{\tau} + \tau')(T_{ln} + T_{m\bar{m}}) - 2\kappa' T_{ll} + \bar{\kappa}T_{nn} + 2\bar{\sigma}T_{n\bar{m}} + 2\sigma' T_{lm}] . \tag{5.32$^{*\prime}$}$$

All other operations lead to one of the above equations.

Commutation Relations

In addition to these fundamental equations we need the commutation relations, which correspond to the Ricci-identities, for GHP operators acting on (p,q) quantities:

$$\text{þ}\text{þ}' - \text{þ}'\text{þ} = (\bar{\tau} - \tau')\eth + (\tau - \bar{\tau}')\eth' - p(\kappa\kappa' - \tau\tau' + \Psi_2) - q(\bar{\kappa}\bar{\kappa}' - \bar{\tau}\bar{\tau}' + \bar{\Psi}_2)\ , \quad (5.33)$$

$$\eth\eth' - \eth'\eth = (\bar{\rho}' - \rho')\text{þ} + (\rho - \bar{\rho})\text{þ}' + p(\rho\rho' - \sigma\sigma' + \Psi_2) - q(\bar{\rho}\bar{\rho}' - \bar{\sigma}\bar{\sigma}' + \bar{\Psi}_2), \quad -(5.33)^*$$

$$\text{þ}\eth - \eth\text{þ} = \bar{\rho}\eth + \sigma\eth' - \bar{\tau}'\text{þ} - \kappa\text{þ}' - p(\rho'\kappa - \tau'\sigma + \Psi_1) - q(\bar{\sigma}'\bar{\kappa} - \bar{\rho}\bar{\tau}')\ , \quad (5.34)$$

$$\text{þ}'\eth' - \eth'\text{þ}' = \bar{\rho}'\eth' + \sigma'\eth - \bar{\tau}\text{þ}' - \kappa'\text{þ} + p(\rho\kappa' - \tau\sigma' + \Psi_3) + q(\bar{\sigma}\bar{\kappa}' - \bar{\rho}'\bar{\tau})\ , \quad (5.34)'$$

$$\text{þ}\eth' - \eth'\text{þ} = \rho\eth' + \bar{\sigma}\eth - \tau'\text{þ} - \bar{\kappa}\text{þ}' - p(\sigma'\kappa - \rho\tau') - q(\bar{\rho}'\bar{\kappa} - \bar{\tau}'\bar{\sigma} + \bar{\Psi}_1), \quad \overline{(5.34)}$$

$$\text{þ}'\eth - \eth\text{þ}' = \rho'\eth + \bar{\sigma}'\eth' - \tau\text{þ}' - \bar{\kappa}'\text{þ} + p(\sigma\kappa' - \rho'\tau) + q(\bar{\rho}\bar{\kappa}' - \bar{\tau}\bar{\sigma}' + \bar{\Psi}_3)\ . \quad \overline{(5.34)}'$$

Table III, which follows from the definition of the spin coefficients will be useful for calculations performed later.

Table III

∇	$n^a\nabla l_a$	$m^a\nabla l_a$	$\bar{m}^a\nabla n_a$	$m^a\nabla\bar{m}_a$	$\frac{1}{2}(n^a\nabla l_a - \bar{m}^a\nabla m_a)$
D	$\varepsilon + \bar{\varepsilon}$	κ	σ'	$\varepsilon - \bar{\varepsilon}$	$-\beta' = \alpha$
Δ	$-(\varepsilon' + \bar{\varepsilon}')$	τ	ρ'	$\bar{\varepsilon}' - \varepsilon'$	β
δ	$\beta - \bar{\beta}'$	σ	κ'	$\beta + \bar{\beta}'$	$-\varepsilon' = \gamma$
$\bar{\delta}$	$\bar{\beta} - \beta'$	ρ	τ'	$-(\bar{\beta} + \beta')$	ε

Table III is equivalent to the following equations:

$$
\begin{aligned}
\text{þ}\, l_a &= -\bar{\kappa}\, m_a - \kappa\, \bar{m}_a , & \eth\, l_a &= -\bar{\rho}\, m_a - \sigma\, \bar{m}_a , \\
\text{þ}\, n_a &= -\tau' m_a - \bar{\tau}' \bar{m}_a , & \eth\, n_a &= -\rho' m_a - \bar{\sigma}' \bar{m}_a , \\
\text{þ}\, m_a &= -\bar{\tau}' l_a - \kappa\, n_a , & \eth\, m_a &= -\bar{\sigma}' l_a - \sigma\, n_a , \\
\text{þ}\, \bar{m}_a &= -\tau' l_a - \bar{\kappa}\, n_a , & \eth\, \bar{m}_a &= -\rho\, l_a - \bar{\rho}\, n_a , \\
\text{þ}'\, l_a &= -\bar{\tau}\, m_a - \tau\, \bar{m}_a , & \eth'\, l_a &= -\bar{\sigma}\, m_a - \rho\, \bar{m}_a , \\
\text{þ}'\, n_a &= -\kappa' m_a - \bar{\kappa}' \bar{m}_a , & \eth'\, n_a &= -\sigma' m_a - \bar{\rho}' \bar{m}_a , \\
\text{þ}'\, m_a &= -\bar{\kappa}' l_a - \tau\, n_a , & \eth'\, m_a &= -\bar{\rho}' l_a - \rho\, n_a , \\
\text{þ}'\, \bar{m}_a &= -\kappa' l_a - \bar{\tau}\, n_a , & \eth'\, \bar{m}_a &= -\sigma' l_a - \bar{\sigma}\, n_a .
\end{aligned}
\tag{5.34}
$$

Clearly, the last three sets of equations in (5.34) may be obtained from the first set by operating with *, ' and *', respectively. From Table III one easily also derives

$$
\begin{aligned}
\nabla^a l_a &= \epsilon + \bar{\epsilon} - \rho - \bar{\rho} , & \nabla^a m_a &= \beta' + \bar{\beta}' - \tau - \bar{\tau}' , \\
\nabla^a n_a &= \epsilon' + \bar{\epsilon}' - \rho' - \bar{\rho}' , & \nabla^a \bar{m}_a &= \beta' + \bar{\beta}' - \tau' - \bar{\tau} .
\end{aligned}
\tag{5.35}
$$

5.10 Tetrad Gauge Transformations

By introducing a tetrad, one introduces another type of gauge arbitrariness in $\mathcal{M}$, namely tetrad gauge (TG) transformations. These can be decomposed into null rotations about l and n, respectively, and boosts in the l-n plane combined with rotations in the orthogonal m-$\bar{m}$ plane. These transformations form three Abelian subgroups of the homogeneous Lorentz group (Janis & Newman 1965) and are given by

$$
\text{(i)} \quad
\begin{aligned}
\tilde{l} &= l , \\
\tilde{m} &= m + a\, l , \\
\tilde{n} &= n + a\, \bar{m} + \bar{a}\, m + a\bar{a}\, l ;
\end{aligned}
\tag{5.36a}
$$

$$\text{(ii)}\quad \begin{aligned} \tilde{n} &= n \\ \tilde{m} &= m + bn \\ \tilde{l} &= l + b\bar{m} + \bar{b}m + b\bar{b}n \ , \end{aligned} \tag{5.36b}$$

$$\text{(iii)}\quad \begin{aligned} \tilde{l} &= r\,l \\ \tilde{n} &= r^{-1}n \\ \tilde{m} &= e^{i\theta}m \ , \end{aligned} \tag{5.36c}$$

where $a,b \in \mathbb{C}$ and $r,\theta \in \mathbb{R}$. Having chosen a principal tetrad, so that perturbations of Ψ_0 and Ψ_4 are PG invariant, the question then arises under what conditions perturbations of the "radiative parts" Ψ_0 and Ψ_4, φ_0 and φ_2 of the gravitational and electromagnetic fields can be TG invariant. To answer this question it suffices to know which quantities are invariant under infinitesimal versions of Eqs. (5.36). For infinitesimal values of $a,b,r = 1+\varepsilon c$ of order ε, (5.36) becomes, up to order ε^2

$$\text{(i)}\quad \begin{aligned} \tilde{l} &= l \\ \tilde{m} &= m + \varepsilon a l \\ \tilde{n} &= n + \varepsilon(a\bar{m} + \bar{a}m) \ , \end{aligned} \qquad \text{(ii)}\quad \begin{aligned} \tilde{l} &= l + \varepsilon(b\bar{m} + \bar{b}m) \\ \tilde{m} &= m + \varepsilon bn \\ \tilde{n} &= n \ , \end{aligned} \tag{5.37}$$

$$\text{(iii)}\quad \begin{aligned} \tilde{l} &= l + \varepsilon c l \\ \tilde{n} &= n - \varepsilon c l \\ \tilde{m} &= m + \varepsilon i\theta m \ . \end{aligned}$$

Let us first compute the behaviour of the tetrad components of the fields and sources as well as the spin coefficients under transformations (5.36), then linearize the results (Janis & Newman 1965; Ludwig 1969, 1974). Note that in (5.37), a has type $(0,-2)$, b has tpye $(2,0)$ and r,θ are both of type $(0,0)$. For null rotations about l these transformation laws are

$$\text{(i)}\quad \tilde{\varphi}_0 = \varphi_0\,, \qquad \tilde{\varphi}_1 = \varphi_1 + \bar{a}\varphi_0\,, \qquad \tilde{\varphi}_2 = \varphi_2 + 2\bar{a}\varphi_1 + \bar{a}^2\varphi_0\,,$$

$$\tilde{\Psi}_0 = \Psi_0\,, \qquad \tilde{\Psi}_1 = \Psi_1 + \bar{a}\Psi_0\,, \qquad \tilde{\Psi}_2 = \Psi_2 + 2\bar{a}\Psi_1 + \bar{a}^2\Psi_0\,,$$

$$\tilde{\Psi}_3 = \Psi_3 + 3\bar{a}\Psi_2 + 3\bar{a}^2\Psi_1 + \bar{a}^3\Psi_0\,, \tag{5.38a}$$

$$\tilde{\Psi}_4 = \Psi_4 + 4\bar{a}\Psi_3 + 6\bar{a}^2\Psi_2 + 4\bar{a}^3\Psi_1 + \bar{a}^4\Psi_0\,;$$

$$\tilde{\kappa} = \kappa\,, \qquad \tilde{\rho} = \rho + \bar{a}\kappa\,, \qquad \tilde{\sigma} = \sigma + a\kappa\,,$$

$$\tilde{\tau} = \tau + \bar{a}\sigma + a\rho + a\bar{a}\kappa\,,$$

$$\tilde{\tau}' = \tau' - \text{þ}\bar{a} - \bar{a}^2\kappa\,, \tag{5.38b}$$

$$\tilde{\rho}' = \rho' + a\tau' - \eth\bar{a} - a\,\text{þ}\bar{a} - \bar{a}^2\sigma - a\bar{a}^2\kappa\,,$$

$$\tilde{\sigma}' = \sigma' + \bar{a}\tau' - \eth'\bar{a} - \bar{a}\,\text{þ}\bar{a} - \bar{a}^2\rho - \bar{a}^3\kappa\,,$$

$$\tilde{\kappa}' = \kappa' + a\sigma' + \bar{a}\rho' - \text{þ}'\bar{a} - a\eth'\bar{a} - \bar{a}\eth\bar{a} - a\bar{a}\,\text{þ}\bar{a} + {}$$
$$+ a\bar{a}\tau' - \bar{a}^2\tau - a\bar{a}^2\rho - \bar{a}^3\sigma - a\bar{a}^3\kappa\,;$$

$$\tilde{T}_{\ell\ell} = T_{\ell\ell}\,, \qquad \tilde{T}_{\ell m} = T_{\ell m} + a T_{\ell\ell}\,,$$

$$\tilde{T}_{mm} = T_{mm} + 2a T_{\ell m} + a^2 T_{\ell\ell}\,,$$

$$\tilde{T}_{\ell n} + \tilde{T}_{m\bar{m}} = T_{\ell n} + T_{m\bar{m}} + 2a T_{\ell\bar{m}} + 2\bar{a} T_{\ell m} + 2a\bar{a} T_{\ell\ell}\,,$$

$$\tilde{T}_{nm} = T_{nm} + a(T_{\ell n} + T_{m\bar{m}}) + \bar{a} T_{mm} + a^2 T_{\ell\bar{m}} + {}$$
$$+ 2a\bar{a} T_{\ell m} + a^2\bar{a} T_{\ell\ell}\,, \tag{5.38c}$$

$$\tilde{T}_{nn} = T_{nn} + 2a T_{n\bar{m}} + 2\bar{a} T_{nm} + 2a\bar{a}(T_{\ell n} + T_{m\bar{m}}) + {}$$
$$+ a^2 T_{\bar{m}\bar{m}} + \bar{a}^2 T_{mm} + 2a^2\bar{a} T_{\ell\bar{m}} + 2a\bar{a}^2 T_{\ell m} + a^2\bar{a}^2 T_{\ell\ell}\,.$$

The transformation law for null rotations about n are just the primed version of Eqs. (5.38) provided one sets $a' = \bar{b}$ and $\varphi'_n = -\varphi_{2-n}$ $(n=0,1,2)$, $\Psi'_m = \Psi_{4-m}$ $(m=0,\ldots,4)$:

(ii) $$\tilde{\varphi}_2 = \varphi_2 \,, \qquad \tilde{\varphi}_1 = \varphi_1 + b\,\varphi_2 \,, \qquad \tilde{\varphi}_0 = \varphi_0 + 2b\,\varphi_1 + b^2\varphi_2 \,,$$

$$\tilde{\Psi}_4 = \Psi_4 \,, \qquad \tilde{\Psi}_3 = \Psi_3 + b\,\Psi_4 \,, \qquad \tilde{\Psi}_2 = \Psi_2 + 2b\,\Psi_3 + b^2\Psi_4 \,,$$

$$\tilde{\Psi}_1 = \Psi_1 + 3b\,\Psi_2 + 3b^2\Psi_3 + b^3\Psi_4 \,, \tag{5.39a}$$

$$\tilde{\Psi}_0 = \Psi_0 + 4b\,\Psi_1 + 6b^2\Psi_2 + 4b^3\Psi_3 + b^4\Psi_4 \;;$$

$$\tilde{\rho} = \rho + \bar{b}\tau - \eth' b - \bar{b}\,\text{þ}' b - b^2\sigma' - b^2\bar{b}\,\kappa' \,,$$

$$\tilde{\sigma} = \sigma + b\tau - \eth b - b\,\text{þ}' b - b^2\rho' - b^3\kappa \,,$$

$$\tilde{\tau} = \tau - \text{þ}' b - b^2\kappa' \,,$$

$$\tilde{\kappa} = \kappa + \bar{b}\sigma + b\rho - \text{þ} b - \bar{b}\,\eth b - b\,\eth' b + b\bar{b}\tau - b\bar{b}\,\text{þ}' b - \tag{5.39b}$$

$$- b^2\tau' - b^2\bar{b}\rho' - b^3\sigma' - b^3\bar{b}\,\kappa' \,,$$

$$\tilde{\rho}' = \rho' + b\kappa' \,,$$

$$\tilde{\sigma}' = \sigma' + \bar{b}\kappa' \,,$$

$$\tilde{\tau}' = \tau' + b\sigma' + \bar{b}\rho' + b\bar{b}\,\kappa' \,,$$

$$\tilde{\kappa}' = \kappa \;;$$

$$\tilde{T}_{nn} = T_{nn} \,, \qquad \tilde{T}_{n\bar{m}} = T_{n\bar{m}} + \bar{b}\,T_{nn} \,,$$

$$\tilde{T}_{\bar{m}\bar{m}} = T_{\bar{m}\bar{m}} + 2\bar{b}\,T_{n\bar{m}} + \bar{b}^2 T_{nn} \,,$$

$$\tilde{T}_{\ell n} + \tilde{T}_{m\bar{m}} = T_{\ell n} + T_{m\bar{m}} + 2\bar{b}\,T_{nm} + 2b\,T_{n\bar{m}} + 2b\bar{b}\,T_{nn} \,, \tag{5.39c}$$

$$\tilde{T}_{\ell\bar{m}} = T_{\ell\bar{m}} + b\,T_{\bar{m}\bar{m}} + \bar{b}\,(T_{\ell n} + T_{m\bar{m}}) + \bar{b}^2 T_{nm} + 2b\bar{b}\,T_{n\bar{m}} + b\bar{b}^2\,T_{nn} \,,$$

$$\tilde{T}_{\ell\ell} = T_{\ell\ell} + 2\bar{b}\,T_{\ell m} + 2b\,T_{\ell\bar{m}} + 2b\bar{b}\,(T_{\ell n} + T_{m\bar{m}}) + \bar{b}^2 T_{mm} +$$

$$+ b^2 T_{\bar{m}\bar{m}} + 2\bar{b}^2 b\,T_{nm} + 2b^2\bar{b}\,T_{n\bar{m}} + b^2\bar{b}^2\,T_{nn} \,.$$

Finally for boosts in the ℓ -n plane and rotations in the m-$\bar{m}$ plane one gets with (5.36c)

(iii) $\tilde{\varphi}_0 = r e^{i\theta} \varphi_0 , \quad \tilde{\varphi}_1 = \varphi_1 , \quad \tilde{\varphi}_2 = r^{-1} e^{-i\theta} \varphi_2 ,$

$$\tilde{\Psi}_0 = r^2 e^{i2\theta} \Psi_0 , \quad \tilde{\Psi}_1 = r e^{i\theta} \Psi_1 , \quad \tilde{\Psi}_2 = \Psi_2 ,$$

$$\tilde{\Psi}_3 = r^{-1} e^{-i\theta} \Psi_3 , \quad \tilde{\Psi}_4 = r^{-2} e^{-2i\theta} \Psi_4 \quad ; \tag{5.40a}$$

$$\tilde{\kappa} = r^2 e^{i\theta} \kappa , \quad \tilde{\sigma} = r e^{2i\theta} \sigma , \quad \tilde{\tau} = e^{i\theta} \tau ,$$

$$\tilde{\kappa}' = r^{-2} e^{-i\theta} \kappa' , \quad \tilde{\sigma}' = r^{-1} e^{-2i\theta} \sigma' , \quad \tilde{\tau}' = e^{-i\theta} \tau' ,$$

$$\tilde{\rho} = r \rho \quad , \quad \tilde{\rho}' = r^{-1} \rho' \quad ; \tag{5.40b}$$

$$\tilde{T}_{\ell\ell} = r^2 T_{\ell\ell} \quad , \quad \tilde{T}_{\ell n} + \tilde{T}_{m\bar{m}} = T_{\ell n} + T_{m\bar{m}} \quad ,$$

$$\tilde{T}_{\ell m} = r e^{i\theta} T_{\ell m} , \quad \tilde{T}_{nm} = r^{-1} e^{i\theta} T_{nm} \quad , \tag{5.40c}$$

$$\tilde{T}_{mm} = e^{2i\theta} T_{mm} , \quad \tilde{T}_{nn} = r^{-2} T_{nn} \quad .$$

Now linearize the above transformations according to (5.37) up to order ε^2 :

(i) $\tilde{\varphi}_n = \varphi_n + \varepsilon n \bar{a} \varphi_{n-1} \qquad (n = 0, 1, 2 ; \ \varphi_{-1} = 0)$

$$\tilde{\Psi}_m = \Psi_m + \varepsilon m \bar{a} \Psi_{m-1} \qquad (m = 0, \ldots, 4 ; \ \Psi_{-1} = 0)$$

$$\tilde{\rho} = \rho + \varepsilon \bar{a} \kappa , \quad \tilde{\rho}' = \rho' + \varepsilon (a \tau' - \eth \bar{a}) ,$$

$$\tilde{\sigma} = \sigma + \varepsilon a \kappa , \quad \tilde{\sigma}' = \sigma' + \varepsilon (\bar{a} \tau' - \eth' \bar{a}) , \tag{5.41}$$

$$\tilde{\tau} = \tau + \varepsilon (a \rho + \bar{a} \sigma) , \quad \tilde{\tau}' = \tau' - \varepsilon \text{Þ} \bar{a} ,$$

$$\tilde{\kappa} = \kappa \quad , \quad \tilde{\kappa}' = \kappa' + \varepsilon (a \sigma' + \bar{a} \rho' - \text{Þ}' \bar{a}) \ ;$$

(ii)
$$\tilde{\varphi}_n = \varphi_n + \varepsilon(2-n)\varphi_{n+1} \qquad (n=0,1,2\,;\ \varphi_3 = 0)$$
$$\tilde{\Psi}_m = \Psi_m + \varepsilon(4-m)\Psi_{m+1} \qquad (m=0,\dots,4\,;\ \Psi_5 = 0)$$
$$\tilde{\rho} = \rho + \varepsilon(\bar{b}\tau - \gamma' b)\,, \qquad \tilde{\rho}' = \rho' + \varepsilon b\kappa'\,,$$
$$\tilde{\sigma} = \sigma + \varepsilon(b\tau - \gamma b)\,, \qquad \tilde{\sigma}' = \sigma' + \varepsilon\bar{b}\kappa'\,, \qquad (5.42)$$
$$\tilde{\tau} = \tau - \varepsilon \text{þ} b\,, \qquad \tilde{\tau}' = \tau' + \varepsilon(\bar{b}\rho' + b\sigma')\,,$$
$$\tilde{\kappa} = \kappa + \varepsilon(\bar{b}\sigma + b\rho - \text{þ} b)\,, \quad \tilde{\kappa}' = \kappa\ ;$$

(iii)
$$\tilde{\varphi}_n = \varphi_n + \varepsilon(2-n)(c+i\theta)\varphi_n \qquad (n=0,1,2\,;\ c = r-1)$$
$$\tilde{\Psi}_m = \Psi_m + \varepsilon(4-m)(c+i\theta)\Psi_m \qquad (m=0,\dots,4)$$
$$\tilde{\kappa} = \kappa + 2\varepsilon(c+i\theta)\kappa\,, \qquad \tilde{\kappa}' = \kappa - 2\varepsilon(c+i\theta)\kappa'\,,$$
$$\tilde{\sigma} = \sigma + \varepsilon(c+i\theta)\sigma\,, \qquad \tilde{\sigma}' = \sigma' - \varepsilon(c-2i\theta)\sigma'\,, \qquad (5.43)$$
$$\tilde{\rho} = \rho + \varepsilon c\rho\,, \qquad \tilde{\rho}' = \rho' - \varepsilon c\rho'\,,$$
$$\tilde{\tau} = \tau + \varepsilon i\theta\tau\,, \qquad \tilde{\tau}' = \tau' - \varepsilon i\theta\tau\,.$$

It should be noted that in studying TG invariant perturbations, in Eqs. (5.41-43), terms involving ε may be replaced by their background values. For example, the first of Eqs. (5.41) becomes $\tilde{\varphi}_n = \varphi_n + \varepsilon n\bar{a}\varphi^{\circ}_{n-1}$ In order that there exist at least one gauge invariant quantity, the unperturbed spacetime must be algebraically special. To see this recall the above remark by Sachs (p.50) and take, e.g., $\rho = 0$. Then if ρ is to be TG invariant, Eqs. (5.41 & 42) demand that $\kappa^{\circ} = \tau^{\circ} = 0$, and from the vacuum field equation (5.28) it then follows that $\sigma^{\circ} = 0$. Within an algebraically special spacetime one easily finds TG invariant quantities which may or may not involve spin coefficients. Take for example $2\Psi^{\circ}_3\varphi_2 - \varphi^{\circ}_1\Psi_4$ or $2\Psi^{\circ}_1\varphi_0 - \varphi^{\circ}_1\Psi_0$; they are both TG invariant.

If, however, one demands that the perturbations of the radiative fields be TG invariant in addition to being PG invariant, then, according to Eqs. (5.41-43) this will be true if and only if the following unperturbed quantities vanish in the background, i.e.

$$\Psi^{\circ}_0 = \Psi^{\circ}_1 = \Psi^{\circ}_3 = \Psi^{\circ}_4 = \varphi^{\circ}_1 = \varphi^{\circ}_2 = \varphi^{\circ}_0 = 0 \quad (\Psi^{\circ}_2 \neq 0)\,. \qquad (5.44)$$

The perturbations then are also PG invariant by the remark of Sachs. Thus we have the result that Ψ_0 and Ψ_4 (and φ_0, φ_2) are PG and TG invariant if and only if (5.44) holds. This, in turn, implies that ℓ and n are tangent to the repeated principle null directions of the unperturbed Weyl tensor which is, therefore, of type $\{22\}$.

Although for suitably chosen tetrads all algebraically special spacetimes contain PG and TG invariant quantities (e.g. Ψ_0 in type $\{211\}$) it is only in type $\{22\}$ that both Ψ_0 and Ψ_4 are PG and TG invariant. They are, in fact, the radiation fields on $\mathcal{J}^-$ and $\mathcal{J}^+$, respectively. These measure the "in- and outgoing" radiation.

5.11 Metric Perturbations Revisited

Now let us return to the question of the relation between curvature and metric perturbations. The analogous question in Electrodynamics is the following: given a field tensor F, does there always exist a potential A? The answer is yes (locally), because Maxwell's equations dF = 0 are just the integrability conditions for the existence of A. The same thing happens in the linearized theory of gravity, but the proof used there does not go over on any obvious way to the more general situations treated here.

For the gravitational and electromagnetic case Chrzanowski (1975) has constructed electromagnetic and metric perturbations from solutions of the respective decoupled and separated NP equations. His construction hinges upon the assumed existence of so-called asymptotically factorized Green's functions. Although strong plausibility arguments can be given, their existence has, however, not been proved yet except for scalar radiation (Chrzanowski & Misner 1974).

In view of the discussions in this chapter let us quote a remark by Ivor & Johanna R. Robinson (1970):
"There are some advantages in approaching the linear approximation to general relativity through the field rather than the potential: geometry ends where gauge transformations begin."

In spite of this, a perturbation can only be considered as acceptable if at least the existence of a metric perturbation is ascertained, irrespective of its gauge-dependence.

VI. DECOUPLED & SEPARATED PERTURBATION EQUATIONS

6.1 Introduction

This chapter gives an account of the treatment of the perturbation equations when the background geometry is of type $\{22\}$, and in particular, when it is a black hole. Using the GHP formalism of Chap. V, wave equations are derived for scalar, electromagnetic and gravitational perturbations, which yield decoupled equations for the radiative field components. Various standard coordinate systems are introduced for Kerr spacetimes and ordinary differential equations are obtained from these decoupled equations by separation of variables. The boundary conditions as implied by the peeling theorem for asymptotically flat spacetimes are discussed for the radial & angular coordinate functions of the separated equations.

Furthermore, various properties of the radial & angular eigenfunctions and eigenvalues are presented, namely interrelations for different spin and expansions of spin-weighted spheroidal harmonics in the limit of low and high frequencies.

6.2 The Decoupling of the Perturbation Equations

Decoupling means in this connection that there exists a differential equation for each of φ_o, φ_2, Ψ_o, Ψ_4, each equation involving only one of these field components. Since only perturbations of type $\{22\}$ spacetimes are considered and since the unperturbed quantities vanish, the perturbed quantities will be denoted by just φ_o, φ_2, Ψ_o, Ψ_4, respectively. Teukolsky (1972) has shown that in the Kerr geometry decoupling of these quantities is possible; Stewart & Walker (1973,1974) have generalized this result and shown that decoupling is possible only if the spacetime is of type $\{22\}$ and non-accelerating (this excludes the types III A and III B of Kinnersley's 1969 classification). The homogeneous decoupled perturbation equations were first rederived within the GHP formalism by Stewart (1972).

In a principal tetrad in a type $\{22\}$ vacuum spacetime the condition of TG invariance of the radiation fields, (5.44), required that

$$\varphi_o = \varphi_1 = \varphi_2 = \Psi_o = \Psi_1 = \Psi_3 = \Psi_4 = 0 \quad (\Psi_2 \neq 0) . \tag{6.1a}$$

The Goldberg-Sachs theorem (cf., e.g., Pirani 1964) then implies

$$\kappa = \sigma = \kappa' = \sigma' = 0. \tag{6.1b}$$

This is trivially seen by inserting Eq. (6.1a) into the vacuum version of (5.31)-(5.31)$^{*\prime}$.

6.3 Scalar Perturbations

For a weak scalar perturbation ϕ of an arbitrary vaccum spacetime its action back on the spacetime via its energy-stress tensor is ignored because it is of second order in ϕ. Thus, ϕ, which is a (0,0) quantity, obeys the covariant inhomogeneous wave equation on the background with a scalar source T, namely

$$\Box\phi = g^{ab}\nabla_a\nabla_b\phi = 2\left[\ell^{(a}n^{b)} - m^{(a}\bar{m}^{b)}\right]\nabla_a\nabla_b\phi = 4\pi T. \tag{6.2}$$

This can also be written as

$$(A_o + A_o^*)\phi = 4\pi T, \qquad \text{where } A_o = (\text{þ}' - \bar{\rho}')\text{þ} - \rho\,\text{þ}' . \tag{6.3}$$

No simplification is achieved by specializing to type $\{22\}$. To become better acquainted with the GHP formalism the derivation of Eq. (6.3) is given in detail. Let us first evaluate

$$\ell^a n^b \nabla_a \nabla_b = (\ell^a \nabla_a)(n^b \nabla_b) - \ell^a (\nabla_a n^b) \nabla_b = D\Delta - (Dn^b)\nabla_b \ .$$

The term Dn^b is given by

$$Dn^b = -(\varepsilon + \bar{\varepsilon}) n^b - \tau' m^b - \bar{\tau}' \bar{m}^b$$

as can be seen from Eq. (5.34). Hence

$$\ell^a n^b \nabla_a \nabla_b = D\Delta + (\varepsilon + \bar{\varepsilon})\Delta + \tau'\delta + \bar{\tau}'\bar{\delta} \ .$$

Similarly, (or by priming the last equation) the other term is

$$n^a \ell^b \nabla_a \nabla_b = \Delta D + (\varepsilon' + \bar{\varepsilon}')D + \bar{\tau}\delta + \tau\bar{\delta} \ ,$$

from which it follows that

$$\begin{aligned} 2\ell^{(a}n^{b)}\nabla_a\nabla_b &= D\Delta + \Delta D + (\varepsilon' + \bar{\varepsilon}')D + (\varepsilon + \bar{\varepsilon})\Delta + (\bar{\tau} + \tau')\delta + (\tau + \bar{\tau}')\bar{\delta} \\ &= \text{Þ}'\text{Þ} + \text{Þ}\text{Þ}' + (\bar{\tau} + \tau')\eth + (\tau + \bar{\tau}')\eth' \\ &= 2(\text{Þ}'\text{Þ} + \bar{\tau}\eth + \tau\eth') \ . \end{aligned}$$

The first step follows from the fact that the operator $\text{Þ}'$ in $\text{Þ}'\text{Þ}\phi = \text{Þ}'(\text{Þ}\phi)$ acts on the (1,1) quantity $\text{Þ}\phi$. According to (5.23) one then has $\text{Þ}' = \Delta + \varepsilon' + \bar{\varepsilon}'$. In the second step $\text{Þ}\text{Þ}'$ is eliminated with the aid of commutation relation (5.33). Clearly, $\Box$ is a star invariant operator. Thus, starring the last equation geves the other half of $\Box$ in Eq. (6.3).

6.4 Electromagnetic Perturbations

As in the scalar case, it is necessary for the decoupling of the radiative field that the perturbed field vanish. For, even though the electromagnetic field enters the energy-stress-tensor quadratically, the perturbation couples to spacetime via the Einstein-Maxwell equations even in first order, if the electromagnetic field does not vanish in the background. Equations analogous to Eq. (6.3) may be obtained from Maxwell's equations in a type $\{22\}$ spacetime as follows. Insert relations (6.1) into Eq. (5.26). Operate on this equation, which consists of quantities of type (1,1), with $(\eth - 2\tau - \bar{\tau}')$. Similarly, to Eq. $(5.26)^*$, which contains quantities of type (1,-1), apply the operator $(\text{þ} - 2\rho - \bar{\rho})$ and subtract. In the resulting equation the term involving φ_1 may be eliminated. This is achieved by means of commutation relation (5.34) for $[\text{þ}, \eth]$ and by using the "field equations" (5.27), $(5.27)^*$ to get rid of the terms $\eth\rho$ and $\text{þ}\tau$. The following decoupled equation for φ_o is left:

$$(A_1 + A_1^*)\ \varphi_o = 2\pi\,(B_1 + B_1^*)\ , \tag{6.4}$$

where

$$A_1 = (\text{þ} - 2\rho - \bar{\rho})(\text{þ}' - \rho')\ , \qquad B_1 = (\eth - 2\tau - \bar{\tau}')\,J_\ell\ .$$

The perturbation equation for φ_2 is simply the prime of (6.4):

$$(A_1' + A_2^{*\prime})\ \varphi_2 = 2\pi\,(B_1' + B_1^{*\prime})\ . \tag{6.5}$$

In the above derivation all quantities but φ_o & φ_2 were evaluated for the background metric, as was done for the scalar wave equation. Therefore, Eqs. (6.3-5) are already linearized about the background and all terms contained in A_o, A_1 and B_1 are evaluated in the unperturbed spacetime.

6.5 Gravitational Perturbations

The linearization prescription in the gravitational case is different. We write each quantity as the sum of an unperturbed and a perturbing part. The unperturbed parts of $\sigma, \kappa, \sigma', \kappa', \Psi_o, \Psi_1, \Psi_3, \Psi_4$ and the test-particle source T_{ab} vanish identically by means of condition (6.1), which is the only case considered here. Hence all terms involving these expressions are automatically at least of first order smallness once the perturbation is made. As a consequence the coefficients of the Ψ's can be replaced by their unperturbed values, i.e.

terms which are quadratic in quantities of first order are neglected.

To arrive at the decoupled equations for Ψ_0 & Ψ_4, start with the Bianchi identities (5.31), (5.31)* , apply $\eth$ to both of them and subtract. With the help of the commutator (5.34) the second derivative of Ψ_1 is eliminated and field equation (5.27) is used to get rid of derivatives of spin coefficients. Of all terms acting on Ψ_2 only $3\Psi_2\Psi_0$ is of first order, all others being second order. Hence the result is

$$(A_2 + A_2^*)\Psi_0 = 4\pi\,(B_2 + B_2^*) \quad , \tag{6.6}$$

where

$$A_2 = (\text{Þ} - 4\rho - \bar{\rho})(\text{Þ}' - \rho') - \tfrac{3}{2}\Psi_2 \quad ,$$

$$B_2 = (\eth - 4\tau - \bar{\tau}')\left[(\text{Þ} - 2\bar{\rho})T_{lm} - (\eth - \bar{\tau}')T_{ll}\right] .$$

The perturbation equation for Ψ_4 is obtained by priming (6.6):

$$(A_2' + A_2^{*\prime}) = 4\pi\,(B_2' + B_2^{*\prime}) \ . \tag{6.7}$$

6.6 Decoupled Master Equation

One may summarize these three decoupled equations (6.3,4,6) in a "master equation" as

$$(A_s' + A_s^{*\prime})\,\Psi_{2s} = 4\pi\,(B_s' + B_s^{*\prime}), \tag{6.8}$$

where for $s = 0,-1,-2$, Ψ_{2s} is $\Psi_0 = \phi, \Psi_{-2} = \Psi_2, \Psi_{-4} = \Psi_4$, respectively. The operator A_s' and the sources B_s' are given by

$$A'_0 = (\text{þ}' - \bar{\rho}')\text{þ} - \rho\,\text{þ}' , \qquad B'_0 = T,$$

$$A'_1 = (\text{þ}' - 2\rho' - \bar{\rho}')(\text{þ} - \rho) , \qquad B'_1 = -\tfrac{1}{2}(\text{þ}' - 2\rho' - \bar{\rho}')\, J_{\bar{m}} ,$$

$$A'_2 = (\text{þ}' - 4\rho' - \bar{\rho}')(\text{þ} - \rho) + \tfrac{3}{2}\Psi_2 , \qquad B'_2 = (\text{þ}' - 4\rho' - \bar{\rho}')[(\eth' - 2\bar{\tau})T_{n\bar{m}} - (\text{þ}' - \bar{\rho}')T_{\bar{m}\bar{m}}] .$$

The operators A'_s can also be written compactly as

$$A'_s = \text{þ}'\text{þ} - (2s\rho' + \bar{\rho}')\text{þ} - \rho\,\text{þ}' + 2s\rho\rho' - s(s - \tfrac{3}{2})\Psi_2 . \qquad (6.9)$$

The corresponding equation for the field Ψ_{2s} with positive spin-weight is the prime of (6.8), namely

$$(A_s + A_s^*)\Psi_{2s} = 4\pi(B_s + B_s^*) , \qquad (6.10)$$

where for $S = 0,1,2$, Ψ_{2s} stands for $\Psi_0 = \phi$, $\Psi_{+2} = \varphi_0$, $\Psi_{+4} = \Psi_0$, respectively. The sources are given by

$$B_0 = T ,$$

$$B_1 = -\tfrac{1}{2}(\text{þ} - 2\rho - \bar{\rho})\, J_m ,$$

$$B_2 = (\text{þ} - 4\rho - \bar{\rho})[(\text{þ}' - 2\bar{\rho}')T_{n\bar{m}} - (\eth' - \bar{\tau})T_{nn}] ,$$

and the operators A_s by

$$A_0 = (\text{þ} - \bar{\rho})\text{þ}' - \rho'\text{þ} ,$$

$$A_1 = (\text{þ} - 2\rho - \bar{\rho})(\text{þ}' - \rho') ,$$

$$A_2 = (\text{þ} - 4\rho - \bar{\rho})(\text{þ}' - \rho') - \tfrac{3}{2}\Psi_2 ,$$

or, equivalently, by

$$A_s = \text{þ}'\,\text{þ} - \bar{\rho}'\,\text{þ} - (2s+1)\,\rho\,\text{þ}' - s(s+\tfrac{1}{2})\Psi_2 \quad . \tag{6.11}$$

For a black hole background the last form of the operator, namely the one involving $\text{þ}'\text{þ}$ instead of $\text{þ}\,\text{þ}'$, is preferable since þ assumes a simpler form than $\text{þ}'$ when expressed in one of the standard coordinate systems.

For the geometry of a rotating black hole with charge Q, i.e. a Kerr-Newman black hole, where in addition to Eqs. (6.17) $\varphi_1^0 = Q\rho^2$, an analogous decoupling program has not been carried out. In fact, from the work done using the Reissner-Nordstrøm geometry (Moncrief 1974 a,b,c; Zerilli 1974, Chitre et al. 1973) one would expect that only for some combination of electromagnetic & gravitational quantities could one obtain decoupled equations. As was seen in Sec. 5.10, tetrad gauge invariant quantities of this kind are easily found. The problem is to find the right one. Assuming a "mildly" charged Kerr-Newman black hole (i.e. neglecting terms quadratic in Q) Chitre (1975) finds from Maxwell's equations a decoupled & separated wave equation for the quantitiy $\rho^{-2}\varphi_2 + \frac{2}{3}\frac{Q}{M}\rho^{-3}\Psi_3$. However, the general problem has not been solved yet.

6.7 Coordinate Systems, Spin-Coefficients & GHP Operators

The Kerr solution was first expressed in Schwarzschild-like coordinates by Boyer & Lindquist (1967), here called BL coordinates. For radiation problems it is sometimes more useful to employ analogous of retarded or advanced "null" coordinates (Carter 1968, Stewart & Walker 1973, MTW). BL, retarded and advanced coordinates cover different patches of the maximally extended Kerr spacetime in the usual Kruskal picture) as shown in Fig. 7.

(a) BL coordinates (r,t,θ,φ) (generalized Schwarzschild coordinates) cover the XXXX patch. The metric in these coordinates is given in Eq. (3.17). The principal NP tetrad for this metric was given by Boyer & Lindquist (1967) $(\Sigma = r^2+a^2\cos^2\theta,\ \Delta_K = r^2-2Mr+a^2)$ as

$$\ell^a = (r^2+a^2, \Delta_K, 0, a)/\Delta_K\,, \quad n^a = (r^2+a^2, -\Delta_K, 0, a)/2\Sigma\,,$$
$$m^a = (ia\sin\theta, 0, 1, i/\sin\theta)/\sqrt{2}(r+ia\cos\theta)\,. \tag{6.12a}$$

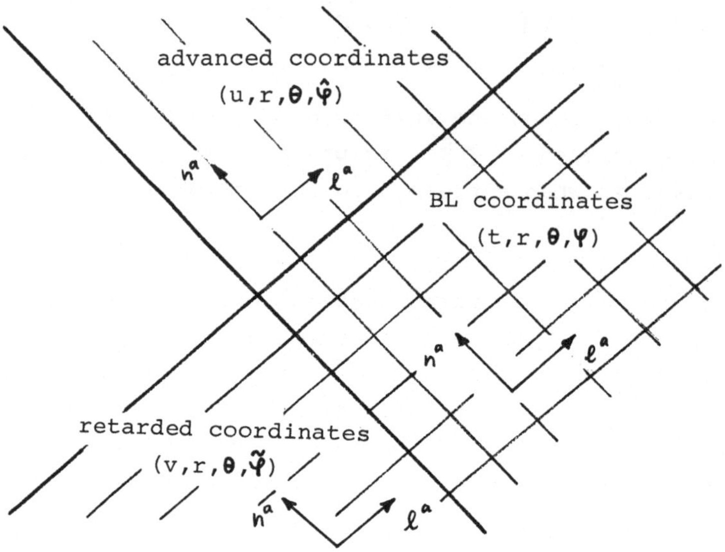

Figure 7. Coordinate Systems for the Kerr Geometry.

This tetrad is singular at the future event horizon where the BL coordinates themselves become singular, but regular on the past horizon. From this tetrad another tetrad can be generated which is regular on the future event horizon but singular on the past event horizon. This tetrad is obtained by interchanging horizons with the aid of the reflection symmetry $(t,\varphi) \rightarrow (-t,-\varphi)$ of the Kerr geometry, is given by the future pointing vectors

$$\ell^a = (r^2+a^2, -\Delta_K, 0, a)/\Delta_K, \quad n^a = (r^2+a^2, \Delta_K, 0, a)/2\Sigma, \tag{6.12b}$$

$$m^a = -(ia\sin\theta, 0, -1, i/\sin\theta)/\sqrt{2}(r+ia\cos\theta).$$

Since BL coordinates degenerate at the future event horizon advanced "null" coordinates are more useful there.

(b) Advanced null coordinates $(v,r,\theta,\hat{\varphi})$ (ingoing Kerr coordinates) cover the \\\ patch. They may be obtained from BL coordinates by means of the transformation

$$dv = dt + \frac{r^2+a^2}{\Delta_K} dr \equiv dt + dr^* ,$$
$$d\hat{\varphi} = d\varphi + \frac{a}{\Delta_K} dr \equiv d\varphi + \frac{a}{r^2+a^2} dr^* , \tag{6.13}$$

with r, θ remaining unchanged. Eqs. (6.13) a new radial coordinate r^* was introduced that will be of use later. A principal tetrad in these coordinates obtained from (6.12a) which is regular at the future horizon is given by

$$\ell^a = (r^2+a^2, \frac{\Delta_K}{2}, 0, a)/\Sigma, \quad n^a = (0,-1,0,0) ,$$
$$m^a = (ia \sin\theta, 0, 1, i/\sin\theta)/\sqrt{2}\,(r+ia \cos\theta). \tag{6.14}$$

(c) Retarded null coordinates $(u, r, \theta, \tilde{\varphi})$ (outgoing Kerr coordinates) cover the /// patch in Fig. 7. They are related to BL coordinates by

$$du = dt - \frac{r^2+a^2}{\Delta_K} dr \equiv dt - dr^* ,$$
$$d\tilde{\varphi} = d\varphi - \frac{a}{\Delta_K} dr \equiv d\varphi - \frac{a}{r^2+a^2} dr^* , \tag{6.15}$$

with r, θ unchanged. The principal tetrad for retarded coordinates, namely

$$\ell^a = (0,1,0,0), \qquad n^a = (r^2+a^2, -\frac{\Delta_K}{2}, 0, a)/\Sigma,$$
$$m^a = (ia \sin\theta, 0, 1, i/\sin\theta)/\sqrt{2}\,(r+ia \cos\theta), \tag{6.16}$$

was given by Kinnersley (1969). The Kerr metric in retarded coordinates reads

$$ds^2 = -\Sigma^{-1}(\Delta_K - a^2\sin^2\theta)du^2 - 2dudr + \Sigma d\theta^2 - 4a\Sigma^{-1}Mr \sin^2\theta du d\tilde{\varphi}$$
$$+ 2a \sin^2\theta \, dr d\tilde{\varphi} + \Sigma^{-1}\left[(r^2+a^2)^2\sin^2\theta - \Delta_K a^2 \sin^4\theta\right] d\tilde{\varphi}^2 .$$

In Eq. (6.16) the non-shearing null geodesics to which ℓ^a is tangent is parametrized with affine parameter r (as it is the case for n^a in (6.14)). The spin coefficients are only functions of r and θ and hence unchanged under transformations (6.13&15). The nonzero ones are (Kinnersley 1969)

$$\begin{aligned} \rho &= -(r - ia\cos\theta)^{-1}, & \tau &= ia\rho\bar{\rho}\sin\theta/\sqrt{2}, \quad \tau' = -ia\rho^2\sin\theta/\sqrt{2}, \\ \rho' &= -\rho^2\bar{\rho}\Delta_K/2, & \varepsilon' &= \rho' - [\text{illegible}](r-M)/2, \quad \beta = -\bar{\rho}\cot\theta/2\sqrt{2}, \qquad (6.17) \\ \beta' &= \bar{\beta} + \tau', & \Psi_2 &= M\rho^3. \end{aligned}$$

These expressions hold in BL, advanced, and retarded coordinates. The GHP operators (5.23) can now be expressed in these coordinates by using Eqs. (6.12,14,16 & 17). When operating on a spin and boos weighted scalar (they can of course be applied to vectors and tensors as well) they become

$$\begin{aligned} \text{(a)} \quad \text{þ} &= \Delta_K^{-1}\left[(r^2+a^2)\nabla_t + \Delta_K\nabla_r + a\nabla_\varphi\right], \\ \text{þ}' &= (2\Sigma)^{-1}\left[(r^2+a^2)\nabla_t - \Delta_K\nabla_r + a\nabla_\varphi + p(M-r-\Delta_K\rho) + q(M-r-\Delta_K\bar{\rho})\right], \\ \eth &= -2^{-1/2}\bar{\rho}\left[ia\sin\theta\,\nabla_t + \nabla_\theta + \frac{i}{\sin\theta}\nabla_\varphi - \frac{p-q}{2}\cot\theta - q\bar{\rho}\,ia\sin\theta\right], \\ \eth' &= -2^{-1/2}\rho\left[ia\sin\theta\,\nabla_t + \nabla_\theta - \frac{i}{\sin\theta}\nabla_\varphi + \frac{p-q}{2}\cot\theta + p\rho\,ia\sin\theta\right]. \end{aligned} \qquad (6.18)$$

As another example, in retarded coordinates these GHP operators are

$$\begin{aligned} \text{(c)} \quad \text{þ} &= \nabla_r, \\ \text{þ}' &= (2\Sigma)^{-1}\Big[2(r^2+a^2)\nabla_u - \Delta_K\nabla_r + 2a\nabla_{\tilde{\varphi}} + (p+q)(M-r) + \\ &\qquad + \Delta_K\Sigma^{-1}\left\{(p+q)r + (p-q)ia\cos\theta\right\}\Big], \\ \eth &= -2^{-1/2}\bar{\rho}\left[ia\sin\theta\,\nabla_u + \nabla_\theta + \frac{i}{\sin\theta}\nabla_{\tilde{\varphi}} - \frac{p-q}{2}\cot\theta - q\bar{\rho}\,ia\sin\theta\right], \\ \eth' &= -2^{-1/2}\rho\left[-ia\sin\theta\,\nabla_u + \nabla_\theta - \frac{i}{\sin\theta}\nabla_{\tilde{\varphi}} + \frac{p-q}{2}\cot\theta + p\rho\,ia\sin\theta\right]. \end{aligned} \qquad (6.19)$$

From now on mostly BL coordinates are used. For this a word of justification is required. In general, BL coordinates are not always appropriate for radiation problems. As can be seen from the spacetime picture in Fig. 8, the lines $t = \text{const}$ cross the world line of an observer (with $r = \text{const}$) far away from the radiating source in a diverging manner for large times ($t \to \infty$). BL-time t is an improper measure of arrival times of signals near $\mathscr{I}^+$. For a distant observer receiving them red (or blue) shifts will enter the expressions for radiation due to this coordinate effect. The appropriate natural measure of time for outgoing radiation, i.e. near $\mathscr{I}^+$, is the retarded time u for which no such divergence effect occurs. However, for our purposes we are seeking absolute magnitudes of neither intensities nor radiated frequencies; everything may be scaled adequately for the situations considered later. In this sense, for radiation going out to $\mathscr{I}^+$ BL coordinates in fact lead to simpler formulae than others.

6.8 Separation of the Master Equation

Separability of the decoupled perturbation equations was demonstrated by Teukolsky (1972) for the Kerr geometry using BL coordinates. Again, Stewart & Walker (1974) showed that this was only possible for those spacetimes of type $\{22\}$ that are not accelerating. In the decoupled master equation (6.10) the fields are of type $(2s,0)$. Hence, before applying the second-order operators of Eq. (6.11) to this equation, note first that $\eth\Psi_{2s}$ is of type $(2s+1,1)$ and $\bar{\eth}\Psi_{2s}$ is of type $(2s+1,-1)$. With the aid of these operators (6.18) the decoupled master equation becomes in BL coordinates

$$\Big\{ \Delta_K^{-s}\, \nabla_r(\Delta_K^{s+1}\nabla_r) + \frac{1}{\sin\theta}\,\nabla_\theta(\sin\theta\,\nabla_\theta) + \Big[\frac{1}{\sin^2\theta} - \frac{a^2}{\Delta_K}\Big]\nabla_\varphi^2 + $$
$$ + 2s\Big[\frac{a(r-M)}{\Delta_K} + \frac{i\cos\theta}{\sin^2\theta}\Big]\nabla_\varphi - \frac{4\,Mar}{\Delta_K}\nabla_t\nabla_\varphi + 2s\Big[\frac{M(r^2-a^2)}{\Delta_K} - r - ia\cos\theta\Big]\nabla_t - $$
$$ - \Big[\frac{(r^2+a^2)^2}{\Delta_K} - a^2\sin^2\theta\Big]\nabla_t^2\Big\}\,\Psi = -4\pi\,\Sigma\,T\,. \tag{6.20}$$

An interesting alternative derivation of Eq. (6.20) has been presented by Ryan (1974). For $s = 0$ this equation was first separated by Brill et al. (1972); for $s = 1$ Cohen & Kegeles (1974) gave another derivation using the method of Debye-potentials suitably generalized for curved spacetimes. If one sets $s = \pm\frac{1}{2}$ then this equation also becomes the wave equation for a neutrino field as was shown by Teukolsky (1974). Eq. (6.20) is, in fact, the coordinate form of the decoupled master equation (6.8) for the field quantities with negative spin weight, provided Ψ and T are chosen appropriately as given in Table IV.

Table IV. Field & Sources for the Master Equation

field Ψ	spin-weight s	source T
${}_0\Omega = \phi$	0	${}_0T = T$
${}_1\Omega = \varphi_0$	+1	${}_1T = B_1 + B_1^*$
${}_{-1}\Omega = \rho^{-2}\varphi_2$	-1	${}_{-1}T = \rho^{-2}(B_1' + B_1^{*\prime})$
${}_2\Omega = \Psi_0$	+2	${}_2T = B_2 + B_2^*$
${}_{-2}\Omega = \rho^{-4}\Psi_4$	-2	${}_{-2}T = \rho^{-4}(B_2' + B_2^{*\prime})$

Eq. (6.20) may be solved by separation of variables. Putting

$$ {}_s\Omega(t,r,\theta,\varphi) = \sum\!\!\!\!\!\!\int \; {}_sR_{\ell m\omega}(r) \; {}_sZ^{\omega}_{\ell m}(\theta,\varphi)\; e^{-i\omega t}\,, \qquad (6.21)$$

where $\sum\!\!\!\!\!\!\int = \sum_{\ell=0}^{\infty}\sum_{m=-\ell}^{\ell}\int_{-\infty}^{\infty} d\omega$, and (t,r,θ,φ) are BL coordinates, leads to ordinary differential equations for each variable. For ${}_sR(r)$ one obtains

$$\left\{\Delta_K^{-s}\frac{d}{dr}\left(\Delta_K^{s+1}\frac{d}{dr}\right) + \frac{K^2 - 2is(r-M)}{\Delta_K} + 4ir\omega s - {}_s\lambda\right\}{}_sR(r) = -T, \qquad (6.22)$$

where ${}_s\lambda$ is a separation constant and $K \equiv (r^2+a^2)\omega - am$. The source term T will be discussed later. The analytic structure of Eq. (6.22) was investigated by Hartle & Wilkins (1974).

The angular eigenfunctions ${}_sZ^{\omega}_{\ell m}$ are given by

$$ {}_sZ^{\omega}_{\ell m}(\theta,\varphi) = \left[\frac{2\ell+1}{4\pi}\,\frac{(\ell-m)!}{(\ell+m)!}\right]^{1/2} {}_sS_{\ell m}(-a\omega,\theta)\,e^{im\varphi} \tag{6.23} $$

and satisfy the normalization conditions

$$ \int d\Omega\; {}_s\bar{Z}^{\omega}_{\ell' m'}(\theta,\varphi)\; {}_sZ^{\omega}_{\ell m}(\theta,\varphi) = \delta_{\ell'\ell}\,\delta_{m'm} \; . \tag{6.24} $$

The functions ${}_sS(-a\omega,\theta)$ are solutions of

$$ \left\{\frac{1}{\sin\theta}\frac{d}{d\theta}\left(\sin\theta\frac{d}{d\theta}\right) + a^2\omega^2\cos^2\theta - 2a\omega s\cos\theta - \frac{(m+s\cos\theta)^2}{\sin^2\theta} + {}_sA_{\ell m}+s\right\}{}_sS(\theta)=0 \tag{6.25} $$

The constant ${}_sA_{\ell m}$ is another separation constant related to ${}_s\lambda$ by ${}_s\lambda = {}_sA + a^2\omega^2 - 2a\omega m$.

The decoupled equations separate as well in retarded & advanced coordinates. Generally, they separate in all systems related to BL coordinates by

$$ \tilde{t} = t+f_1(r)+f_2(\theta), \quad \tilde{\varphi} = \varphi + g_1(r)+g_2(\theta), \quad \tilde{r} = h(r), \; \tilde{\theta} = J(\theta). $$

For advanced coordinates Eq. (6.21) is replaced by

$$ {}_s\Omega(v,r,\theta,\hat{\varphi}) = \sum\!\!\!\!\!\!\int \; {}_sR_{\ell m\omega}(r)\,{}_sZ^{\omega}_{\ell m}(\theta,\hat{\varphi})\,e^{-i\omega v} \tag{6.26a} $$

and instead of (6.22) one obtains the radial equation

$$ \left\{\Delta_K^{-s}\frac{d}{dr}\left(\Delta_K^{s+1}\frac{d}{dr}\right) - 2iK\frac{d}{dr} - \frac{4is(r-M)}{\Delta_K} - 2i\omega r(2s+1) - {}_s\lambda\right\}{}_sR(r) = -T. \tag{6.26b} $$

In retarded coordinates one starts with

$$ {}_s\Omega(u,r,\theta,\tilde{\varphi}) = \sum\!\!\!\!\!\!\int {}_sR_{\ell m\omega}(r)\ {}_sZ^{\omega}_{\ell m}(\theta,\tilde{\varphi})e^{-i\omega u} \tag{6.27a} $$

and obtains

$$ \left\{\Delta_K^{-s}\frac{d}{dr}\left(\Delta_K^{s+1}\frac{d}{dr}\right) + iK\frac{d}{dr} + 2(2s+1)i\omega r - {}_s\lambda\right\}{}_sR(r) = -T \ . \tag{6.27b} $$

If one works with the tetrad given by (6.12b) then ${}_sR(r)$ has to be replaced by ${}_s\bar{R}(r)$ in Eqs. (6.22,26,27).

The radial equation (6.22) has regular singular points at the two horizons r_+, r_- (which coalesce in the extreme Kerr case). In addition, it has an irregular singular point at infinity. Except for special cases ($a\omega \ll 1$ and $a\omega \gg 1$) no analytic treatment of this equation is possible. Hence for the general case the equation has to be integrated numerically. For the purpose of investigating the stability this was done by Press & Teukolsky (1974).

For subsequent purposes it is useful to cast Eq. (6.22) into the form of a one-dimensional Schrödinger-type equation. This can be done for all linear second-order ordinary differential equations with polynomial coefficients. First, introducing a new radial coordinate r^* given by

$$ r^* = \int dr(r^2+a^2)/\Delta_K \ , $$

leads to

$$ \left\{\frac{d^2}{dr^{*2}} + 2\,G\frac{d}{dr^*} + \frac{K^2-2is(r-M)K + \Delta_K(4ir\omega s-{}_s\lambda)}{(r^2+a^2)^2}\right\}{}_sR(r) = T, \tag{6.28} $$

where $G \equiv s(r-M)/(r^2+a^2) + r\,\Delta_K/(r^2+a^2)^2$. The new radial coordinate varies from $-\infty$ to $+\infty$ when r varies between r_+ and ∞. Introducing a new radial function

$$ {}_sR(r) = \exp\left(-\int G dr^*\right){}_su(r) = \Delta_K^{-s/2}(r^2+a^2)^{-1/2}\,{}_su(r) \tag{6.29} $$

eliminates the first order derivative in Eq. (6.28) and yields

$$\left[- \frac{d^2}{dr^{*2}} + {}_sV(r) \right] {}_su(r) = \Delta_K^{1+s/2} (r^2+a^2)^{-3/2} T , \qquad (6.30)$$

where

$${}_sV(r) = - \frac{K^2 - 2is(r-M)K + \Delta_K(4ir\omega - {}_s\lambda)}{(r^2+a^2)^2} + G^2 + \frac{dG}{dr^*}$$

is a complex "effective" potential. The source term T is given by

$$T = \begin{cases} 4\pi \int d\Omega\, dt\, \Sigma\, {}_sT\, {}_s\bar{Z}^{\omega}_{\ell m}(\theta,\varphi)\, e^{i\omega t} & \text{for tetrad (6.12a)}, \\ 4\pi \int d\Omega\, dt\, \Sigma\, {}_s\bar{T}\, {}_sZ^{\omega}_{\ell m}(\theta,\varphi)\, e^{-i\omega t} & \text{for tetrad (6.12b)}, \end{cases} \qquad (6.31)$$

with ${}_sT$ as in Table IV. The above radial and angular differential equations allow the following symmetry relations for the corresponding coordinate eigenfunctions:

$$\begin{aligned} {}_sR_{\ell m\omega}(r) &= (-1)^m\, {}_s\bar{R}_{\ell\,-m\,-\omega}(r) , \\ {}_s\bar{R}_{\ell m\omega}(r) &= (1/\Delta_K)^s\, {}_{-s}R_{\ell m\omega}(r) , \\ {}_s\bar{Z}^{\omega}_{\ell m}(\theta,\varphi) &= (-1)^m\, {}_{-s}Z^{-\omega}_{\ell\,-m}(\theta,\varphi) , \\ {}_sZ^{\omega}_{\ell m}(\pi-\theta, \pi+\varphi) &= (-1)^{\ell}\, {}_{-s}Z^{\omega}_{\ell m}(\theta,\varphi) . \end{aligned} \qquad (6.32)$$

It has been shown by Press & Teukolsky (1974) that there exist operators which provide algebraic relations between quantities of positive and negative spin-weight and their first derivatives. Define angular and radial operators $\mathcal{L}_n$ and $\mathcal{D}$ by

$$\begin{aligned} \mathcal{L}_n &= \nabla_\theta + \frac{m}{\sin\theta} - a\omega\sin\theta + n\cot\theta , \\ \mathcal{L}_n^+ &= \mathcal{L}_n(-\omega,-m) = \nabla_\theta - \frac{m}{\sin\theta} + a\omega\sin\theta + n\cot\theta , \\ \mathcal{D} &= \nabla_r - iK/\Delta_K , \\ \mathcal{D}^+ &= \mathcal{D}(-\omega,-m) = \nabla_r + iK/\Delta_K . \end{aligned} \qquad (6.33)$$

Then from Maxwell's equations one may derive the following relations

$$\mathcal{L}_0 \mathcal{L}_1 \; {}_1S_{\ell m} = B \; {}_{-1}S_{\ell m} \, , \qquad \mathcal{D}\mathcal{D} \; {}_{-1}R_{\ell m\omega} = \frac{1}{2} \; {}_1R_{\ell m\omega} \, ,$$

$$\mathcal{L}_0^+ \mathcal{L}_1^+ \; {}_{-1}S_{\ell m} = B^2 \; {}_1S_{\ell m} \, , \qquad \mathcal{D}^+ \mathcal{D}^+ \Delta_K \; {}_1R_{\ell m\omega} = 2B^2 \Delta_K^{-1} \; {}_{-1}R_{\ell m\omega} . \tag{6.34}$$

Here B is a normalization constant given by

$$B^2 = Q^2 + 4a\omega m - 4a^2\omega^2, \qquad \text{where } Q = {}_sA + s(s+1) + a^2\omega^2 .$$

For a gravitational case similar equations may be derived, namely

$$\begin{aligned}
\mathcal{L}_{-1} \mathcal{L}_0 \mathcal{L}_1 \mathcal{L}_2 \; {}_2S_{\ell m} &= \mathrm{Re}\,[C] \; {}_{-2}S_{\ell m} \, , \\
\mathcal{L}_{-1}^+ \mathcal{L}_0^+ \mathcal{L}_1^+ \mathcal{L}_2^+ \; {}_{-2}S_{\ell m} &= (C^2 - i \,\mathrm{Im}\,[C]) \; {}_2S_{\ell m} \, , \\
\mathcal{D}\mathcal{D}\mathcal{D}\mathcal{D} \; {}_{-2}R_{\ell m\omega} &= \frac{1}{4} \; {}_2R_{\ell m\omega} \, , \\
\mathcal{D}^+ \mathcal{D}^+ \mathcal{D}^+ \mathcal{D}^+ \; {}_2R_{\ell m\omega} &= 4 \, |C|^2 \Delta_K^{-2} \; {}_{-2}R_{\ell m\omega} \, .
\end{aligned} \tag{6.35}$$

Starobinsky & Churilov (1973) have shown that the normalization factor C is given by

$$\begin{aligned}
|C|^2 &= (Q^2 + 4a\omega m - 4a^2\omega^2) \left[(Q-2)^2 + 36a\omega m - 36a^2\omega^2 \right] \\
&\quad + (2Q-1)(96a^2\omega^2 - 48a\omega m) + 144\,\omega^2 (M-a^2) ,
\end{aligned} \tag{6.36}$$

$$\mathrm{Im}[C] = 12\, M\omega, \qquad \mathrm{Re}\,[C] = + (|C|^2 - \mathrm{Im}\,[C])^{1/2} .$$

6.9 Completeness of Eigenfunctions

A problem which arises in relation with the angular and radial equations is the completeness of their eigenfunctions. One defines two kinds of completeness.

Definition 6.1

Let $\{\xi_n\}$ be a countable set of functions with $\xi_n \in \mathbb{L}$, $n \in \mathbb{N}$, where $\mathbb{L}$ is the linear space of all C^∞ complex functions on the unit sphere. Let $(Q_1,Q_2) = \int d\Omega \bar{Q}_1 Q_2$ be the inner product defined on $\mathbb{L}$. Let the norm of Q be defined by $\|Q\| = (Q,Q)^{1/2}$. Then the set $\{\xi_n\}$ is said to be weakly complete if $\forall Q \in \mathbb{L}$, $n \in \mathbb{N}$, $(Q,\xi_n) = 0 \Rightarrow Q = 0$. On the other hand, $\{\xi_n\}$ is said to be strongly complete if $\forall Q \in \mathbb{L}$, $\|Q\|^2 = \sum_{n\in\mathbb{N}} |(Q,\xi_n)|^2$.

Strong completeness implies weak completeness. An arbitrary perturbation can be decomposed into normal modes only if strong completeness holds. For the angular equation, Stewart (1975) was able to establish strong completeness only for the case when $a\omega$ and therefore ω is real. In this case, whether or not strong completeness of the radial eigenfunctions holds, a normal mode expansion is justified. Since strong completeness holds for the φ and t variables as well as for ${}_sS(\theta)$ when ω is real, the set of eigenfunctions $\{{}_sR_{\ell m\omega}(r)\,{}_sS_{\ell m}(\theta)e^{im\varphi}e^{-i\omega t}\}$ is strongly complete even if the set $\{{}_sR_{\ell m\omega}(r)\}$ is only weakly complete (see e.g., Friedman 1965). In general, if a solution of an equation in n variables is sought by separation of variables, then for a decomposition into normal modes to be justified, strong completeness of eigenfunctions of only (n-1) variables is required.

It is a demonstration of strong completeness of the ${}_sS$ for imaginary frequencies which is still missing to complete a rigorous proof of stability of the Kerr metric. For $s=0$ the eigenfunctions are actually strongly complete for all frequencies. However, to prove it for general s seems to constitute an open problem in functional analysis.

6.10 Boundary Conditions

The boundary conditions (Misner 1972, Z'eldovich 1972, Teukolsky 1973) needed to determine solutions of Eq. (6.30) are initial data to be put on some smooth spacelike hypersurface C (a Cauchy surface). If one considers a normal mode expansion of the perturbation as in (6.21), the Cauchy development $D^+(C)$ of C (cf. Hawking & Ellis 1973) of each mode is trivially determined via the $e^{-i\omega t}$ factor as is the φ-dependence. One only needs to solve the equations for the functions ${}_sR(r)$

and ${}_sS(\theta)$ on C to know the complete field throughout $D^+(C)$.

The hypersurfaces $t = \text{const}$ suggested by BL coordinates do not provide an appropriate choice for the Cauchy surface. They either end at the intersection of the future & past horizon or extend into the inside of the collapsing star. This is the case if the exterior of the spacetime considered is that of a black hole formed by gravitational collapse (see Fig. 8). It is tempting to try and take as Cauchy surface a $t = \text{const}$ surface extending only up to the surface of the collapsing body. However, radiation (among other surprises) from the upper remainder of the collapsing star's surface could then reach $\mathcal{J}^+$ without passing through this surface. Consequently, $\mathcal{J}^+$ would not be completely contained in $D^+(C)$. But strong future asymptotic predictability demands that $\mathcal{J}^+$ is entirely contained in $D^+(C)$.

An appropriate Cauchy surface is a smooth hypersurface extending from the future event horizon to spacelike infinity I^o. Such a partial Cauchy surface suffices for our purposes as only the spacetime outside the horizon, the "region of outer communications" (Carter), is of interest here. (C would be a global Cauchy surface for the asymptotically flat region if it extended from I^o across the horizon to the singularity.) In order to construct a partial Cauchy surface one notes that the only coordinate among the ones mentioned here that covers the future event horizon is advanced time v; at I^o this surface may be $t = \text{const}$. Here one requires that

(i) C tends to $t = \text{const}$ as $r \to \infty$ ($r^* \to \infty$),
(ii) C tends to $v = \text{const}$ as $r \to r_+$ ($r^* \to \infty$).

In fact, it is only necessary to demand that C is of finite variation in t and v in the neighbourhood of I^o and r_+, respectively. Eq. (6.30) is regular on C everywhere except at the singular points $r^* = \pm\infty$. Hence the solution is regular except for $r^* = \pm\infty$, where regularity conditions have to be imposed. Due to asymptotic flatness the fields ${}_s\Omega$ have to satisfy the peeling theorem at $\mathcal{J}^+$ (Newman & Penrose 1962). Together with the regularity conditions imposed on the function ${}_sS(\theta)$, the Cauchy data on C have to satisfy

(i) ${}_sS(\theta)$ is finite for $0 \leq \theta \leq \pi$.
(ii) ${}_sR(r)$ is finite for all r if $s \geq 0$; $\varrho^{-2}{}_{-1}R(r)$ and $\varrho^{-4}{}_{-2}R(r)$ are finite for all r if $s \leq 0$.
(iii) At $r \to \infty$, the fields satisfy the peeling theorem.

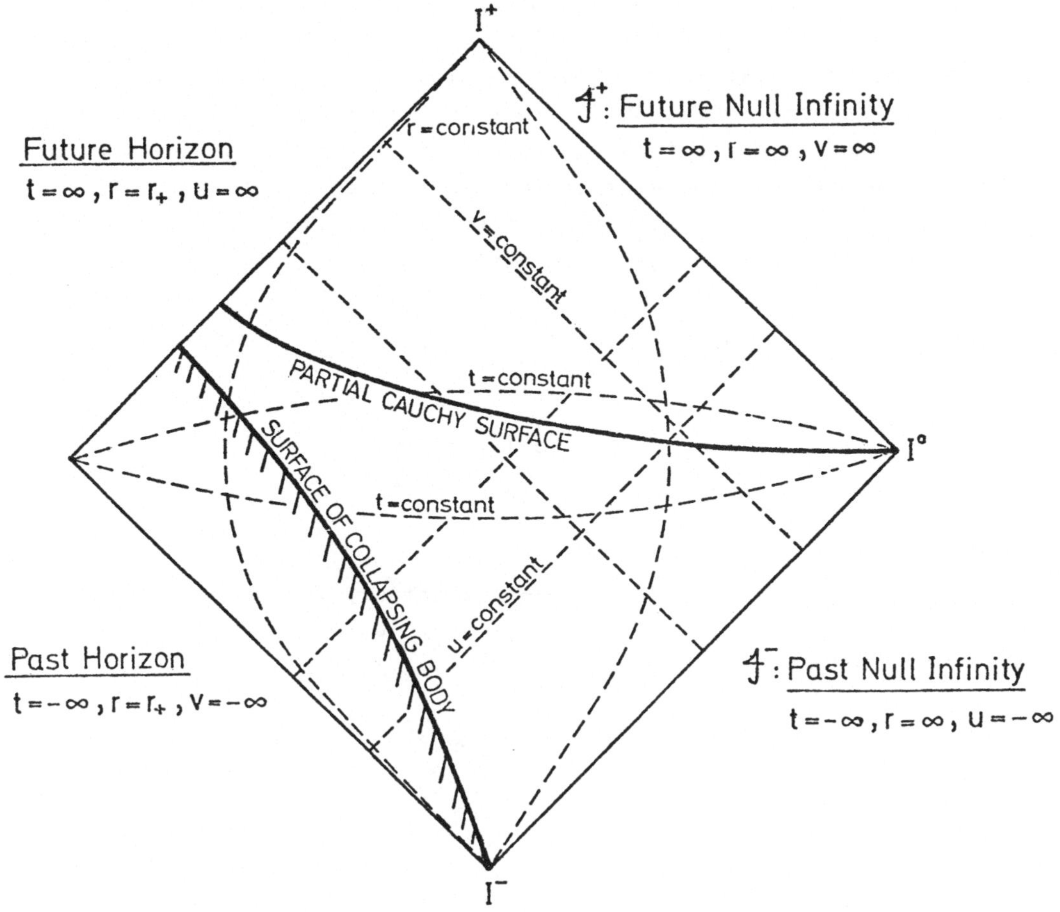

Figure 8. A Penrose diagram (Penrose 1964) for the exterior Kerr spacetime. Dashed lines indicate the surfaces of constant time in BL, retarded and advanced coordinates t,u,v, respectively (from Stewart 1975b).

Condition (i) determines the eigenvalue ${}_sA$, while condition (ii) fixes the eigenvalue ${}_s\lambda$. [In earlier literature boundary conditions were sometimes specified for $r^* \to \pm\infty$, i.e. at I^o and for the entire future event horizon; but this does not constitute a Cauchy problem.] The peeling theorem determines the asymptotic behaviour of the fields near $\mathcal{J}^+$. One obtains the asymptotic behaviour of solutions of Eq. (6.22) by considering the solutions of the asymptotic forms of Eq.(6.30). For $r^* \to \pm\infty$ the latter equations becomes

$$-\frac{d^2{}_su}{dr^{*2}} - \left(k_+^2 + \frac{2ik_+s}{r}\right){}_su = 0, \quad r^* \to +\infty, \tag{6.35a}$$

$$-\frac{d^2{}_su}{dr^{*2}} + \left[k_-^2 - \frac{2is(r_+-M)k_-}{2Mr_+} - \frac{s^2(r_+-M)^2}{(2Mr_+)^2}\right]{}_su = 0, \quad r^* \to -\infty, \tag{6.35b}$$

where $k_+ = \omega, k_- = \omega - m\omega_h$, $\omega_h = \frac{a}{2Mr_+}$. Consider, for example, Eq. (6.35a). Its solutions are ${}_su_{1/2} \sim r^{\pm s} e^{\mp ik_+r^*}$ or

$${}_sR_1 \sim r^{-2s-1}\, e^{i\omega r}, \quad {}_sR_1 \sim r^{-1}\, e^{-i\omega r}. \tag{6.35c}$$

The peeling property can be used to select appropriate boundary conditions for ${}_sR(r)$ near I^o. However, the peeling theorem by Newman & Penrose gives an expansion of the fields in the neighbourhood of $\mathcal{J}^+$ along null geodesics using an affine parameter r_{NP}. On the other hand, the BL coordinate r_{BL} cannot be used for the peeling theorem until one knows the relationship between them. For large values of r_{BL}, i.e. near $\mathcal{J}^\pm$ & I^o, the u,v = const lines ($u = t-r$, $v = t+r$) asymptotically approach Minkowskian null geodesics, and both r_{BL} and r_{NP} approach (up to a factor) the same spherical coordinate r of Minkowski spacetime. Therefore, one proceeds as follows: take some point on C near I^o, study the asymptotic behaviour of the fields from this point in the limits u = const, $r_{BL} \to \infty$ and v = const, $r_{BL} \to \infty$, respectively, and select the fields which are of order $1/r$ near $\mathcal{J}^\pm$. Inserting (6.35c) into (6.21) and using Table IV leads to

$$(\Psi_{2s})_1 \sim r^{-2s-1}\, e^{i\omega r^*}\, e^{-i\omega t} = r^{-2s-1}\, e^{i\omega u},$$

$$(\Psi_{2s})_2 \sim r^{-1}\, e^{-i\omega r^*}\, e^{-i\omega t} = r^{-1}\, e^{-i\omega v}.$$

Now consider the first of the two limits mentioned above. The $e^{-i\omega t}$ factor remains bounded on C for $r \to \infty$, and ${}_sR_1$ blows up exponentially if $\mathrm{Im}(\omega) < 0$. But the fields ${}_s\Omega$ corresponding to ${}_sR_1$ satisfy the peeling theorem if $\mathrm{Im}(\omega) \geq 0$. Hence regularity condition (ii) and the peeling theorem allow only $(\Psi_{2s})_1$ for $u = \mathrm{const}$, $r \to \infty$. Similarly, the limit $v = \mathrm{const}$, $r \to \infty$ reveals that only $(\Psi_{2s})_2$ is allowed near $\mathscr{I}^-$. These solutions may be identified as <u>outgoing</u> and <u>ingoing radiation</u> on $\mathscr{I}^+$ and $\mathscr{I}^-$, respectively. This implies the following radial behaviour for the fields on C near I^o:

$$\phi, \varphi_2, \Psi_4 \sim r^{-1} e^{i\omega r^*}, \qquad \varphi_o \sim r^{-3} e^{i\omega r^*}, \qquad \Psi_o \sim r^{-5} e^{i\omega r^*},$$

for outgoing radiation and

$$\phi, \varphi_o, \Psi_o \sim r^{-1} e^{i\omega r^*}, \qquad \varphi_2 \sim r^{-3} e^{i\omega r^*}, \qquad \Psi_4 \sim r^{-5} e^{i\omega r^*},$$

for ingoing radiation. Similarly one analyzes Eq. (6.35b) and obtains in terms of ${}_sR(r)$,

$$ {}_sR(r) \sim \begin{cases} r^{-2s-1} e^{i\omega r^*} & \text{for } r^* \to \infty \text{ on C,} \\ \Delta_K^{-s} e^{-ik_- r^*} & \text{for } r^* \to -\infty \text{ on C.} \end{cases} \tag{6.36}$$

Table IV (Teukolsky & Press 1974) summarizes all boundary conditions for in- and outgoing radiation with respect to the three standard coordinate systems and the two tetrads given in (6.12 a & b). Let us now define a set of scattering solutions of the homogeneous Eq. (6.30):

$$ {}_s\Omega^{in} = \Delta_k^{-s/2} (r^2+a^2)^{-1/2} \; {}_su^{in}(r) \; {}_sZ(\theta,\varphi) \; e^{-i\omega t}, \tag{6.37a}$$

$$ {}_s\Omega^{up} = \Delta_K^{-s/2} (r^2+a^2)^{-1/2} \; {}_su^{up}(r) \; {}_sZ(\theta,\varphi) \; e^{-i\omega t}, \tag{6.37b}$$

$$ {}_s\Omega^{out} \propto {}_s\bar{\Omega}^{in} , \quad {}_s\Omega^{down} \propto {}_s\bar{\Omega}^{up} . \tag{6.37c}$$

Table V. Asymptotic Solutions for ${}_sR(r)$

	outgoing waves	ingoing waves	outgoing waves	ingoing waves
tetrad (6.12a)				
(t,φ)	$e^{i\omega r^*}/r^{2s+1}$	$e^{-i\omega r^*}/r$	$e^{ik_- r^*}$	$\Delta_K^{-s}\ e^{ik_- r^*}$
$(u,\tilde{\varphi})$	r^{-2s-1}	$e^{-2i\omega r^*}/r$	1	$\Delta_K^{-s}\ e^{-2ik_- r^*}$
$(v,\hat{\varphi})$	$e^{2i\omega r^*}/r^{2s+1}$	r^{-1}	$e^{2ik_- r^*}$	Δ_K^{-s}
tetrad (6.12b)				
(t,φ)	$e^{i\omega r^*}/r$	$e^{-i\omega r^*}/r^{2s+1}$	$\Delta_K^{-s}\ e^{ik_- r^*}$	$e^{-ik_- r^*}$
$(u,\tilde{\varphi})$	r^{-1}	$e^{-2i\omega r^*}/r^{2s+1}$	Δ_K^{-s}	$e^{-2ik_- r^*}$
$(v,\hat{\varphi})$	$e^{2i\omega r^*}/r$	r^{-2s-1}	$\Delta_k^{-s}\ e^{2ik_- r^*}$	1

They satisfy the following boundary conditions: ${}_s\Omega^{in}$ represents a purely incoming wave from $\mathscr{I}^-$ and ${}_s\Omega^{up}$ has as initial state ($v = -\infty$) a wave coming up from the past horizon (see Fig. 9). Solutions "in" and "down", "up" and "out" are related to each other by the conditions that they coincide (in both amplitude and phase) at the future event horizon and at $\mathscr{I}^+$, respectively. The corresponding asymptotic forms for the functions ${}_sR(r)$ can be determined with the help of Table IV and Eq. (6.29). They are

$$
{}_su^{in} \sim \begin{cases} |k_+|^{-1/2}\left[r^s e^{-ik_+ r^*} + S\ r^{-s} e^{ik_+ r^*}\right], & r^* \to \infty, \\ |k_-|^{-1/2}\mathcal{T}\Delta_K^{-s/2}\ e^{ik_- r^*}, & r^* \to -\infty, \end{cases} \tag{6.38a}
$$

$$
{}_su^{up} \sim \begin{cases} |k_+|^{-1/2}\ e^{ik_+ r^*}/r^s, & r^* \to \infty, \\ |k_-|^{-1/2}\left[\mathcal{T}^{-1}\Delta_K^{s/2}\ e^{ik_- r^*} - \overline{(S/\mathcal{T})}\Delta_K^{-s/2}\ e^{ik_- r^*}\right]\frac{k_-}{|k_-|}\frac{k_+}{|k_+|}, & r^* \to -\infty . \end{cases} \tag{6.38b}
$$

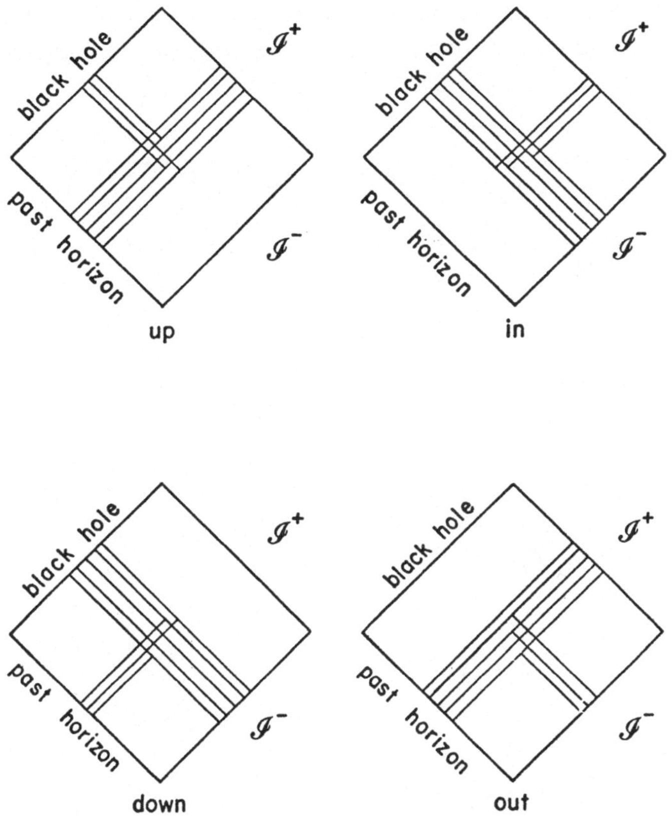

Figure 9. The boundary conditions defining ${}_s\Omega^{in}$, ${}_s\Omega^{out}$, ${}_s\Omega^{up}$ and ${}_s\Omega^{down}$ are illustrated by drawing wave packets built from these fields on Penrose diagrams.The scattering states ${}_s\Omega^{up}$ and ${}_s\Omega^{in}$ are characterized by the behaviour of the incident wave packet: ${}_s\Omega^{up}$ is a wave initially coming "up" from the past horizon and ${}_s\Omega^{in}$ consists of incident "ingoing"radiation. The labels "down" and "out" refer to the charcteristic feature of the outgoing state, namely the fact that the entire wave packet is going "down" the black hole in the former case and "out" to $\mathscr{I}^+$ in the latter (from Chrzanowski & Misner 1974).

where S and $\mathcal{T}$ are reflection and transmission coefficients, respectively. The incoming waves are normalized to unit incident flux. With the aid of (6.38) one can easily show that on the solutions (6.37) have the asymptotic forms behaviour at $\mathscr{I}^+$

$$
\begin{aligned}
{}_s\Omega^{up} &\sim |\omega|^{-1/2}\, r^{2s-1}\, {}_sZ(\theta,\varphi)\, e^{i\omega(r^*-t)} ,\\
{}_s\Omega^{out} &\sim |\omega|^{-1/2}\, r^{-1}\, {}_sZ(\theta,\varphi)\, e^{i\omega(r^*-t)} .
\end{aligned}
\tag{6.38c}
$$

6.10 Radial Green's Function Solution

In this section the radial equation (6.30) is solved by the method of Green's functions. The treatment presented here is based on Chrzanowski (1975). This equation is

$$\left[-\frac{d^2}{dr^{*2}} + {}_sV(r)\right] {}_su_{\ell m\omega}(r) = (r^2+a^2)^{-3/2}\, \Delta_K^{1+s/2} (\Sigma\, {}_sT)_{\ell m\omega} , \tag{6.39}$$

where

$$(\Sigma\, {}_sT)_{\ell m\omega} = 4\pi \int d\Omega\, dt\, \Sigma\, {}_sT\, {}_s\bar{Z}^{\omega}_{\ell m}(\theta,\varphi)\, e^{i\omega t}$$

and the ${}_sT$ are defined in Table IV. Denoting the Green's function by $G(r^*,r_o^*)$, the solution of Eq. (6.39) is given by

$${}_su_{\ell m\omega}(r) = \int dr_o^*\, G(r^*,r_o^*) \left[(r^2+a^2)^{-3/2}\, \Delta_K^{1+s/2} (\Sigma\, {}_sT)_{\ell m\omega}\right]_{r^*=r_o^*} . \tag{6.40}$$

The equation defining G is

$$\left[-\frac{d^2}{dr^{*2}} + {}_sV(r)\right] G(r^*,r_o^*) = \delta(r^*,r_o^*) . \tag{6.41}$$

In order to solve this equation, one takes two linearly independent solutions ${}_su^{in}$ and ${}_su^{up}$ of the homogeneous Eq. (6.39), which satisfy the boundary conditions (6.38). These are then matched at $r^* = r_o^*$. One requires $G \propto {}_su^{up}$ for $r > r_o$ and $G \propto {}_su^{in}$ for

$r^* < r_o^*$. The remaining two arbitrary constants are fixed by the matching conditions $[G] = 0$, $[G'] = -1$ at $r^* = r_o^*$, which is required for a delta-function source. Hence the Green's function G is given by

$$G(r^*, r_o^*) = \frac{i}{2} \frac{\omega}{|\omega|} \begin{cases} {}_s u^{in}(r_o^*)\ {}_s u^{up}(r^*), & r^* > r_o^*, \\ {}_s u^{up}(r_o^*)\ {}_s u^{in}(r^*), & r^* < r_o^*, \end{cases} \qquad (6.42)$$

and has the asymptotic behaviour

$$G(r^*, r_o^*) \sim \frac{i}{2} \frac{\omega}{|\omega|} \begin{cases} |\omega|^{-1/2}\ {}_s u^{in}(r_o)\, r^{-s} e^{i\omega r^*}, & r^* \to \infty, \\ |k_-|^{-1/2}\ {}_s u^{up}(r_o)\ \Delta_K^{-s/2}\ e^{-ik_- r}, & r^* \to -\infty. \end{cases} \qquad (6.43)$$

Inserting (6.43) into (6.40), expressing ${}_s u^{up}$ in terms of ${}_s R^{up}(r)$ for $r^* > r_o^*$ by means of (6.29), and using $d^4x\sqrt{-g} = \Sigma\, d\Omega\, dr dt$ gives

$${}_s R^{up}_{\ell m\omega}(r) = i \frac{\omega}{|\omega|}\ {}_s R^{up}_{\ell m\omega}(r) \int d^4x_o \sqrt{-g}\, \Delta_K^S\ {}_s T_s R^{in}_{\ell m\omega}\ {}_s \bar{Z}^{\omega}_{\ell m\omega} e^{i\omega t}. \qquad (6.44)$$

Then with the aid of Eqs. (6.32) and (6.37) the field becomes (neglecting normalization factors for the moment; see Sec 8.3)

$${}_s\Omega^{up} = \sum\!\!\!\!\!\!\int i \frac{\omega}{|\omega|}\ {}_s\Omega^{up}_{\ell m\omega} \int d^4x_o \sqrt{-g}\ \overline{\left[{}_{-s}R^{out}_{\ell m\omega}\ {}_s Z^{\omega}_{\ell m}\ e^{-i\omega t}\right]}\ {}_s T$$

$$= \sum\!\!\!\!\!\!\int i \frac{\omega}{|\omega|}\ {}_s\Omega^{up}_{\ell m\omega} \left\langle {}_{-s}R^{out}_{\ell m\omega}\ {}_s Z^{\omega}_{\ell m}\ e^{-i\omega t},\ {}_s T \right\rangle, \qquad (6.45)$$

where the inner product is defined by

$$\langle \phi, T \rangle = \int d^4x \sqrt{-g}\ \bar{\phi}\, T. \qquad (6.46)$$

The inner product occurring in Eq. (6.45) is evaluated by partial integration. For this purpose some auxiliary formulae are required showing how to integrate NP quantities by parts. Let U, V, f_1 and f_2 be arbitrary functions of coordinates x^a. Then it can be shown that

($\int \equiv \int d^4x\sqrt{-g}$)

$$\int U\cdot\nabla_1 V = -\int V\cdot\nabla_1 U - \int U\cdot V\,\nabla_a z_1^a \quad , \tag{6.47a}$$

$$\int U\cdot(\nabla_1+f_1)(\nabla_2+f_2)V = \int V\cdot(\nabla_2-f_2+\nabla_a z_2^a)(\nabla_1-f_1+\nabla_a z_1^a)U \quad , \tag{6.47b}$$

where both (∇_1, z_1^a) and (∇_2, z_2^a) stand for one of (D, ℓ^a), (Δ, n^a), (δ, m^a) or $(\bar{\delta}, \bar{m}^a)$. The quantities $\nabla_a z^a$ were calculated in Eqs. (5.35b). In addition, from the field equations (5.26-29) specialized to type $\{22\}$ one can deduce that in Kinnersley's tetrad ($\epsilon = 0$) one has the specific relations

$$D\rho/\rho = \rho \;, \quad \Delta\rho = \rho' \,, \quad \delta\rho/\rho = \tau \,, \quad \bar{\delta}\rho/\rho = \tau' \,. \tag{6.48}$$

The scalar product in (6.45) can now be integrated by parts. The general result ist

$$\langle {}_{-s}R^{out}\; {}_{s}Ze^{-i\omega t}, {}_{s}T \rangle = \langle {}_{s}X_{\alpha}^{out}, S^{\alpha} \rangle , \tag{6.49}$$

where S^{α} is T, J^a, or T^{ab} for $s = 0, \pm 1, \pm 2$. Note that ${}_{-s}R\; {}_{s}Ze^{-i\omega t}$ is of type (o,-2s) since ${}_{s}R$ is of type (s,s) and ${}_{s}Z$ is of type (s,-s). If in the special case of an electromagnetic field ($s = \pm 1$) one inserts ${}_{\pm 1}T$ from Table IV, one obtains

$$\langle {}_{-1}X_a^{out}, J^a \rangle = \int d^4x\sqrt{-g}\left[{}_{-1}\bar{X}_{\ell}\, J_n + {}_{-1}\bar{X}_m\, J_m\right], \tag{6.50a}$$

$$\langle {}_{+1}X_a^{out}, J^a \rangle = \int d^4x\sqrt{-g}\left[-{}_{+1}\bar{X}_n\, J_{\ell} - {}_{+1}\bar{X}_m\, J_m\right], \tag{6.50b}$$

where

$$ {}_{-1}X_a^{out} = \bar{\rho}^{-2}\left\{-m_a(\text{þ} - \bar{\rho}') + n_a(\eth - \bar{\tau}')\right\} {}_{+1}R^{out}\; {}_{-1}Z\; e^{-i\omega t} , \tag{6.51a}$$

$$ {}_{1}X_a^{out} = \left\{-\ell_a(\eth' + \bar{\tau}) + \bar{m}_a(\text{þ} + \bar{\rho})\right\} {}_{-1}R^{out}\; {}_{+1}Z\; e^{-i\omega t} . \tag{6.51b}$$

In order to arrive at (6.51), relations (6.47a & 48) have been used. Because of using relations (6.48), which are not of good type, expressions (6.51)-(6.58) hold only for the tetrad employed here. Similarly, the partial integration required in (6.49) can be worked out for the gravitational case ($s = \pm 2$) with the aid of (6.47b & 48) and the ${}_{\pm 2}T$ taken from Table IV. The result is

$$\langle {}_2X^{out}_{ab}, T^{ab} \rangle = \int d^4x \sqrt{-g} \left[{}_2\bar{X}_{nn} T_{\ell\ell} + {}_2\bar{X}_{mm} T_{mm} - {}_2\bar{X}_{(nm)} T_{\ell m} \right], \qquad (6.52a)$$

$$\langle {}_{-2}X^{out}_{ab}, T^{ab} \rangle = \int d^4x \sqrt{-g} \left[{}_{-2}\bar{X}_{\ell\ell} T_{nn} - {}_{-2}\bar{X}_{(\ell\bar{m})} T_{mm} + {}_{-2}\bar{X}_{\bar{m}\bar{m}} T_{\bar{m}\bar{m}} \right], \qquad (6.52b)$$

where

$$ {}_{-2}X^{out}_{ab} = \bar{\varrho}^{-4} \Big\{ -n_a n_b (\eth - 5\bar{\tau}')(\eth - \bar{\tau}') - m_a m_b (\text{þ}' - 5\bar{\tau}')(\text{þ}' - \bar{\varrho}') + $$

$$ + n_{(a}m_{b)} \left[(\eth - 5\bar{\tau}' + \tau)(\text{þ}' - \varrho') + (\text{þ}' - 5\bar{\varrho}' + \varrho')(\eth - \bar{\tau}') \right] \Big\} \; {}_2R^{out} \; {}_{-2}Z \; e^{i\omega t}, $$

$$(6.53a)$$

$$ {}_2X^{out}_{ab} = \Big\{ - \ell_a \ell_b \; (\eth' - \bar{\tau})(\eth + 3\bar{\tau}) - \bar{m}_a \bar{m}_b (\text{þ} - \bar{\varrho})(\text{þ} + 3\bar{\varrho}) + $$

$$ + \ell_{(a}\bar{m}_{b)} \left[(\text{þ} + \varrho - \bar{\varrho})(\eth' + 3\bar{\tau}) + (\eth' + \tau' - \bar{\tau})(\text{þ} + 3\bar{\varrho}) \right] \Big\} \; {}_{-2}R^{out} \; {}_{+2}Z \; e^{i\omega t}, $$

$$(6.53b)$$

Thus the general result for the field generated by the source ${}_sT$ is

$$\Omega_s = \sum\!\!\!\!\!\!\int \, i \frac{\omega}{|\omega|} \qquad {}_s\Omega^{up}_{\ell m\omega} \langle {}_sX^{out}_{\alpha} , {}_s{}^{\kappa} \rangle . \qquad (6.54)$$

Polarization States

The expressions obtained in Eqs. (6.51 & 53) have still to be decomposed into two independent solutions which constitute the two transverse orthogonal polarization states of the perturbation. The symmetries of the Kerr geometry under the parity operation $\mathcal{P} = (\theta \to \pi - \theta, \varphi \to \varphi + \pi)$ may be used to identify these polarization states by writing

$$A^{out}_a(P=\pm) = {}_{+1}X^{out}_a \pm \mathcal{P}\, {}_{\pm 1}X^{out}_a , \quad h^{out}_{ab}(P=\pm) = {}_{+2}X^{out}_{ab} \pm \mathcal{P}\, {}_{\pm 2}X^{out}_{ab} . \qquad (6.55)$$

The parity operation $\mathcal{P}$ has the following effect on the NP quantities:

$$\mathcal{P}x = x, \ \mathcal{P}y = \bar{y}, \ \mathcal{P}z = -\bar{z}, \tag{6.56}$$

where

$x = \ell_a, n_a, D, \Delta$; $y = \varepsilon, \varrho, \varepsilon', \varrho'$ and c.c. terms ;

$z = m_a, \delta, \tau, \tau', \beta, \beta'$ and c.c. terms.

Thus, for <u>outgoing radiation</u> (s = -1,-2) one has

$$A_a^{out}(P=\pm) = \varrho^{-2}\{\ \}_{+1}R^{out}{}_{-1}Z\ e^{i\omega t} \pm \varrho^{-2}\{c.c.\}_{\pm 1}R^{out}{}_{+1}Z\ e^{-i\omega t}, \tag{6.57a}$$

$$h_{ab}^{out}(P=\pm) = \bar{\varrho}^{-4}\{\ \}_{+2}R^{out}{}_{-2}Z\ e^{i\omega t} \pm \varrho^{-4}\{c.c.\}_{+2}R^{out}{}_{+2}Z\ e^{-i\omega t}, \tag{6.57b}$$

where the curly brackets are defined in Eqs. (6.51a & 53a), respectively. Finally, the individual polarization states are given by taking + or - signs in the (6.57), i.e.

$$A_a^{out}(P=+) = 2\mathrm{Re}\,\mathfrak{f}_{-1}X_a^{out}, \qquad h_{ab}^{out}(P=+) = 2\mathrm{Re}\,\mathfrak{f}_{-2}X_{ab}^{out},$$
$$A_a^{out}(P=-) = 2i\mathrm{Im}\,\mathfrak{f}_{-1}X_a^{out}, \qquad h_{ab}^{out}(P=-) = 2i\mathrm{Im}\,\mathfrak{f}_{-2}X_{ab}^{out}. \tag{6.58}$$

Chrzanowski (1975) shows that expressions A_a and h_{ab} may, in fact, be identified as potentials for the corresponding electromagnetic and gravitational NP perturbations satisfying the "gauges"

$$A_a^{out}n^a = 0 \quad \text{and} \quad h_{ab}^{out}n^b = h^{out}{}_a{}^a = 0.$$

6.11 <u>Spin-Weighted Spherical & Spheroidal Harmonics</u>

If in Eq. (6.25) we set $a\omega = 0$, then a complete and orthonormal set of eigenfunctions of the resulting equation (6.58a) is given by the so-called <u>spin-weighted</u> <u>spherical</u> <u>harmonics</u> ${}_sY_{\ell m}(\theta,\varphi)$, which are defined on the unit sphere. In this case the separation constant ${}_sA_{\ell m}$ assumes the value $(\ell-s)(\ell+s+1)$. The ${}_sY_{\ell m}$ can be generated from ordinary scalar spherical harmonics $Y_{\ell m}(\theta,\varphi) \equiv {}_0Y_{\ell m}(\theta,\varphi)$ by repeatedly operating on these with $\eth, \eth'$ as shown below. These operators are identical to those defined in Eq. (5.23) except for the (in this case purely radial) factor $-\bar{\varsigma}/\sqrt{2} = 1/r\sqrt{2}$.

Let $\mathbb{L}$ be the linear space of all C^∞ complex functions Q of spin-weight s and azimuthal dependence $e^{im\varphi}$ defined on the unit sphere. Then these operators are defined by (Gelfand et al. 1963; Goldberg et al. 1966)

$$\begin{aligned} \eth Q &= -(\sin\theta)^s\left[\nabla_\theta + \frac{i}{\sin\theta}\nabla_\varphi\right]\left\{(\sin\theta)^{-s}Q\right\}, \\ \eth' Q &= -(\sin\theta)^{-s}\left[\nabla_\theta - \frac{i}{\sin\theta}\nabla_\varphi\right]\left\{(\sin\theta)^{s}Q\right\}. \end{aligned} \tag{6.59}$$

The quantities $\eth Q$, $\eth' Q$ have spin-weight $(s+1)$, $(s-1)$, respectively. Hence $\eth'\eth$ leaves the spin-weight of Q unchanged. The functions ${}_sY_{\ell m}$ are by definition

$${}_sY_{\ell m}(\theta,\varphi) = \begin{cases} \left[\dfrac{(\ell-s)!}{(\ell+s)!}\right]^{1/2} \eth^s\, Y_{\ell m}(\theta,\varphi); & 0 \leq s \leq \ell, \\[2ex] \left[\dfrac{(\ell+s)!}{(\ell-s)!}\right]^{1/2} (-\eth')^{-s}\, Y_{\ell m}(\theta,\varphi), & -\ell \leq s \leq 0. \end{cases} \tag{6.60}$$

They are not defined for $|s| > \ell$. The orthonormality and completeness relations are

$$\begin{aligned} \int d\Omega\; {}_s\bar{Y}_{\ell' m'}(\theta,\varphi)\; {}_sY_{\ell m}(\theta,\varphi) &= \delta_{\ell'\ell}\,\delta_{m'm}, \\ \sum_{\ell,m} {}_sY_{\ell m}(\theta',\varphi')\; {}_sY_{\ell m}(\theta,\varphi) &= \delta(\cos\theta - \cos\theta')\,\delta(\varphi-\varphi'). \end{aligned} \tag{6.61}$$

For $a\omega = 0$, Eq. (6.25) reduces to

$$\eth'\eth\; {}_sY_{\ell m} = -(\ell-s)(\ell+s+1)\; {}_sY_{\ell m}, \tag{6.62a}$$

showing that the ${}_sY_{\ell m}$ are eigenfunctions of $\eth'\eth$. Further properties of the spin-weighted spherical harmonics are $(|s| \leq \ell)$ given by

$$ {}_s\bar{Y}_{\ell m} = (-1)^{s+m}\, {}_{-s}Y_{\ell m} \,, \qquad (6.62b)$$

$$ \eth\, {}_sY_{\ell m} = \left[(\ell-s)(\ell+s+1)\right]^{1/2}\, {}_{s+1}Y_{\ell m} \,, \qquad (6.62c)$$

$$ \eth'\, {}_sY_{\ell m} = -\left[(\ell+s)(\ell-s+1)\right]^{1/2}\, {}_{s-1}Y_{\ell m} \,, \qquad (6.62d)$$

$$ \eth\eth'\, {}_sY_{\ell m} = -(\ell+s)(\ell-s+1)\, {}_sY_{\ell m} \,, \qquad (6.62e)$$

$$ \left[\eth'\eth - \eth\eth'\right] Q = 2sQ, \quad \text{for } Q \in \mathbb{L}, \qquad (6.62f)$$

$$ \int d\Omega\, Q_1\, \eth\, Q_2 = -\int d\Omega\, Q_2\, \eth\, Q_1 \,, \qquad (6.62g)$$

$$ \eth'^{p}\, \eth^{p}\, {}_sY_{\ell m} = (-1)^p \frac{(\ell-s)!\,(\ell+s+p)!}{(\ell+s)!\,(\ell-s-p)!}\, {}_sY_{\ell m} \,, \qquad (6.62h)$$

where $d\Omega$ is the line element on the sphere. A useful representation of the ${}_sY_{\ell m}$ is

$$ {}_sY_{\ell m}(\theta,\varphi) = \left[\frac{2\ell+1}{4\pi}\, \frac{(\ell+m)!\,(\ell-m)!}{(\ell+s)!\,(\ell-s)!}\right]^{1/2} (\sin\tfrac{\theta}{2})^{2\ell} \times \qquad (6.63)$$
$$ \times \sum_n \binom{\ell-s}{n} \binom{\ell+s}{n+s-m} (-1)^{\ell-s-n} (\cot\tfrac{\theta}{2})^{2n+s-m}\, e^{im\varphi} \,.$$

Eq. (6.62a) shows that the ${}_sY_{\ell m}$ are eigenfunctions of a singular Sturm-Liouville problem with eigenvalue $(\ell-s)(\ell+s+1)$. On the other hand, for the general Kerr case $(a\omega \neq 0)$, Eq. (6.25) constitutes a 2-parameter eigenvalue problem with eigenvalues $({}_sA_{\ell m}, a\omega)$. But this equation is no Sturm-Liouville differential equation.

The eigenfunctions of Eq. (6.25) are <u>spin-weighted spheroidal harmonics.</u> For the general Kerr case operators $\eth, \eth'$ may be defined as in (5.23); they turn out to be the operators $\mathcal{L}_s^+$, $\mathcal{L}_s$ respectively, as defined in (6.33). Interesting enough, with these operators $\mathcal{L}_s$, $\mathcal{L}_s^+$ the angular wave equation (6.25) <u>cannot</u> be written in a form as simple as in (6.58a), but assumes the more general form

$$\mathcal{L}_{s+1}\ \mathcal{L}_s^+\ {}_sZ^{\omega}_{\ell m} = \left\{ 2a\omega\left[(2s+1)\cos\theta + m\right] - a^2\omega^2 - {}_sA_{\ell m}\right\} {}_sZ^{\omega}_{\ell m}. \tag{6.64}$$

For $a\omega \neq 0$, this is no longer an eigenvalue equation. Therefore, $\mathcal{L}_s$ & $\mathcal{L}_s^+$ are inappropriate for generating the ${}_sZ_{\ell m}$ from scalar spheroidal harmonics $Z_{\ell m} \equiv {}_0Z_{\ell m}$ in a fashion analogous to (6.60). In view of this, it may even be inappropriate to call the eigenfunctions of (6.64) spin-weighted spheroidal harmonics; this name should in fact be given to the eigenfunctions of the "canonical" operator $\mathcal{L}^+_{s+1}\ \mathcal{L}_s$. Eq. (6.25) can be expressed in yet another way with the help of the spherical operator $\eth$:

$$\left[\eth'\eth - 2sa\omega\cos\theta + a^2\omega^2\cos^2\theta\right] {}_sZ^{\omega}_{\ell m} = -{}_sA_{\ell m}\ {}_sZ^{\omega}_{\ell m}. \tag{6.65}$$

In the concluding two sections further properties of spin-weighted spheroidal harmonics are given. An expansion is made in terms of spin-weighted spherical harmonics for the case $a\omega \ll 1$, and the analytic functions they approach in the high-frequency limit ($a\omega \gg 1$) are derived.

6.12 Low-Frequency Expansion of Spin-Weighted Spheroidal Harmonics

In this section the eigenfunctions and eigenvalues of the spheroidal wave equation are examined for the case $a\omega \ll 1$ (Breuer & Ryan, 1975). Letting $x = \cos\theta$ the angular equation (6.25) for the eigenfunctions ${}_sS_{\ell m}(-a\omega, x)$ and eigenvalues ${}_sA_{\ell m}(a\omega)$ becomes

$$\left[\frac{d}{dx}(1-x^2)\frac{d}{dx} + a^2\omega^2x^2 - 2a\omega sx - \frac{(m+sx)^2}{1-x^2} + s\right] {}_sS_{\ell m} = -{}_sA_{\ell m}\ {}_sS_{\ell m}. \tag{6.62}$$

For $a\omega = 0$, Eq. (6.66) reduces to the equation

$$\left[\frac{d}{dx}(1-x^2)\frac{d}{dx} - \frac{(m+sx)^2}{1-x^2} + s\right] {}_sP_{\ell m} = -(\ell-s)(\ell+s+1)\ {}_sP_{\ell m}\ , \tag{6.67}$$

whose solutions are the spin-weighted spherical harmonics ${}_sS_{\ell m}(0,x) \equiv {}_s\mathcal{P}_{\ell m}(x)$. In the case $s = 0$ Eq. (6.63) is the equation for the associated Legendre-polynomials $P_\ell^m(x)$. The left hand side of Eqs. (6.66 & 67) differ by the term $a^2\omega^2x^2 - 2a\omega sx$. For small values of $a\omega$ ($\ll 1$) the influence of this additional term in (6.66) on the solutions and eigenvalues of Eq. (6.67) can be treated as a perturbation. Standard perturbation methods lead to the following expansion for the eigenvalues:

$$
{}_sA_{\ell m} = \begin{cases} (\ell-s)(\ell+s+1) + 2a\omega\dfrac{s^2 m}{\ell(\ell+1)} + O(a^2\omega^2), & s = 0, \\[2ex] (\ell+1) - \dfrac{a^2\omega^2}{2}\left[1 + \dfrac{(2m-1)(2m+1)}{(2\ell-1)(2\ell+3)}\right] + O(a^4\omega^4), & s \neq 0. \end{cases} \tag{6.68}
$$

When $a\omega$ is not infinitesimal, then the perturbation method may be replaced by a continuation method yielding differential equations in $a\omega$ which can be integrated numerically. In fact, Press & Teukolsky (1974) have listed polynomial expansions for ${}_{\pm 2}A_{\ell m}$ for $\ell \leq 6$ and all allowed m up to order $a^7\omega^7$.

Let us next seek an expansion of the spheroidal eigenfunctions ${}_sS_{\ell m}$ in terms of the spherical eigenfunctions ${}_s\mathcal{P}_{\ell m}$ (Breuer & Ryan 1975). Eq. (6.67) has three regular singular points $\pm 1, \infty$ and can therefore be solved by means of hypergeometric functions. The Riemann symbol for Eq. (6.67) is

$$
\left\{ \begin{matrix} +1 & -1 & \infty & \\ \lambda & \mu & \nu & x \\ \lambda' & \mu' & \nu' & \end{matrix} \right\} \tag{6.69}
$$

The elements occurring in this symbol satisfy the algebraic relations

$$
\lambda + \lambda' = 0, \qquad \mu + \mu' = 0, \qquad \nu + \nu' = 1,
$$

$$
\frac{2\lambda\lambda'}{(x-1)^2(x+1)} - \frac{2\mu\mu'}{(x-1)(x+1)^2} + \frac{\nu\nu'}{(x-1)(x+1)} = -\frac{(m+sx)^2}{(x-1)^2(x+1)^2} + \frac{\ell(\ell+1)-s^2}{1-x^2}, \tag{6.70}
$$

which have the solutions

$$\lambda = -(s+m)/2, \qquad \mu = -(s-m)/2, \qquad \nu = -\ell,$$
$$\lambda' = (s+m)/2, \qquad \mu' = (s-m)/2, \qquad \nu' = \ell+1. \tag{6.71}$$

One would like to find Jacobi-polynomials $P_n^{(\alpha,\beta)}(x)$ that are solutions to Eq. (6.67) and regular at $x = \pm 1$. Two cases have to be distinguished. For $|m| \geq |s|$ the only hypergeometric functions that have such a form and are finite at both $x = \pm 1$ are (cf. Magnus et al. 1966)

$$\left(\frac{x-1}{x+1}\right)^{\frac{s+m}{2}} (x+1)^{\ell} \; F\left\{ -(\ell-m), -(\ell-s);\; 1+(s+m);\; \frac{x-1}{x+1} \right\}, \tag{6.72a}$$

$$\left(\frac{x+1}{x-1}\right)^{\frac{s-m}{2}} (x-1)^{\ell} \; F\left\{ -(\ell+m), -(\ell-s);\; 1+(s+m);\; \frac{x+1}{x-1} \right\} . \tag{6.72b}$$

The corresponding solutions of Eq. (6.63) for $|m| \geq |s|$ are hence given by

$${}_sP_{\ell m}(x) = (x-1)^{\frac{m+s}{2}} (x+1)^{\frac{m-s}{2}} P_{\ell-m}^{(m+s,m-s)}(x) \quad , \qquad (m \geq 0) \tag{6.73a}$$

$${}_sP_{\ell m}(x) = (x+1)^{\frac{s-m}{2}} (x-1)^{-\frac{s+m}{2}} P_{\ell+m}^{(-s-m,s-m)}(x) \quad , \qquad (m \leq 0) \tag{6.73b}$$

For $|m| < |s|$ one similarly obtains solutions of Eq. (6.67) of the form

$${}_sP_{\ell m}(x) = (x+1)^{\frac{s+m}{2}} (x-1)^{\frac{s-m}{2}} P_{\ell-s}^{(s+m,s-m)}(x) \quad , \qquad (s > 0) \tag{6.74a}$$

$${}_sP_{\ell m}(x) = (x+1)^{-\frac{s+m}{2}} (x-1)^{-\frac{s-m}{2}} P_{\ell+s}^{(-s-m,-s+m)}(x) \quad , \; (s < 0) \quad . \tag{6.74b}$$

One assumes that for $|m| \geq |s|$ the eigenfunctions ${}_sS_{\ell m}$ can be expanded in terms of the ${}_sP_{\ell m}$ as given in Eq. (6.73). Thus

$$ {}_sS_{\ell m}(x) = (x-1)^{\frac{m+s}{2}} (x+1)^{\frac{m-s}{2}} \sum_{n=m}^{\infty} d^n_{\ell m} \, P^{(m+s,m-s)}_{n-m}(x) , \qquad (m \geqslant 0) \qquad (6.75a) $$

$$ {}_sS_{\ell m}(x) = (x+1)^{\frac{s-m}{2}} (x-1)^{-\frac{s-m}{2}} \sum_{n=-m}^{\infty} d^n_{\ell m} \, P^{(-s-m,s-m)}_{n+m}(x) , \quad (m \leqslant 0) \qquad (6.75b) $$

for some coefficients $d^n_{\ell m}$, which satisfy the five-term recursion relation

$$ a^2\omega^2 \frac{(n+s+2)(n-s+2)(n+s+1)(n-s+1)}{(n+1)(n+2)(2n+3)2n+5)} d^{n+2}_{\ell m} - \frac{2(n+2+1)(n-s+1)}{(n+1)(2n+3)} \left[a\omega s + a^2\omega^2 \frac{sm}{n(n+2)} \right] d^{n+1}_{\ell m} + $$

$$ + \left[{}_sA_{\ell m} + (n-s)(n+s+1) + a^2\omega^2 \left\{ \frac{s^2m^2}{n^2(n+1)^2} + \frac{(n-s)(n+s)(n-m)(n+m)}{n^2(2n-1)(2n+1)} \right. \right. $$

$$ \left. \left. + \frac{(n-m+1)(n+m+1)(n-s+1)(n+s+1)}{(2n+1)(n+1)^3(2n+3)} \right\} + 2a\omega \frac{ms^2}{n(n+1)} \right] d^n_{\ell m} - $$

$$ - \frac{2(n-m)(n+m)}{n(2n-1)} \left[a\omega s + a^2\omega^2 \frac{sm}{(n-1)(n+1)} \right] d^{n-1}_{\ell m} + $$

$$ + a^2\omega^2 \frac{(n-m-1)(n+m+1)(n-m)(n+m)}{(n-1)n(2n-3)(2n-1)} d^{n-2}_{\ell m} = 0. \qquad (6.76) $$

It may be worthwhile to mention that for $s = 0$ the eigenfunctions of (6.67), the associated Legendre polynomials, are related to the $P^{(m,m)}_{n-m}$ in (6.75) through the Gegenbauer polynomials $C^{m+1/2}_{n-m}(x)$. For $|m| < |s|$ the corresponding expansions in terms of the functions ${}_sP_{\ell m}$ given in (6.74) are

$$ {}_sS_{\ell m}(x) = (x+1)^{\frac{s+m}{2}} (x-1)^{\frac{s-m}{2}} \sum_{n=s}^{\infty} \tilde{d}^n_{\ell m} P^{(s+m,s-m)}_{n-s}(x) , \qquad (s \geqslant 0) \qquad (6.77a) $$

$$ {}_sS_{\ell m}(s) = (x+1)^{-\frac{s+m}{2}} (x-1)^{-\frac{s-m}{2}} \sum_{n=-s}^{\infty} \tilde{d}^n_{\ell m} \, P^{(-s-m,-s+m)}_{n+s}(x) , \quad (s \leqslant 0) . \ (6.77b) $$

Here the expansion coefficients $\tilde{d}^n_{\ell m}$ again satisfy a five-term recursion relation, namely

$$a^2\omega^2 \frac{(n-m+1)(n+m+1)(n-m+2)(n+m+2)}{(n+1)(n+2)(2n+3)(2n+5)} \tilde{d}^{n+2}_{\ell m} - \frac{2(n-m+1)(n+m+2)}{(n+1)(2n+3)} \left[a\omega s + a^2\omega^2 \frac{sm}{n(n+1)} \right] \tilde{d}^{n+1}_{\ell m} +$$

$$+ \left[{}_sA_{\ell m} + (n-s)(n+s+1) + a^2\omega^2 \left\{ \frac{s^2 m^2}{n^2(n+1)^2} + \frac{(n-s)(n+s)(n-m)(n+m)}{n^2(2n-1)(2n+1)} + \right.\right.$$

$$\left.\left. + \frac{(n-s+1)(n+s+1)(n-m+1)(n+m+1)}{(2n+1)(2n+3)(n+1)^2} \right\} + 2a\omega \frac{ms^2}{n(n+1)} \right] \tilde{d}^{n}_{\ell m} -$$

$$- \frac{2(n-s)(n+2)}{n(2n-1)} \left[2a\omega s + a^2\omega^2 \frac{sm}{(n-1)(n+2)} \right] \tilde{d}^{n-1}_{\ell m} +$$

$$+ a^2\omega^2 \frac{(n-s-1)(n+s-1)(n-s)(n+s)}{(n-1)n(2n-3)(2n-1)} \tilde{d}^{n-2}_{\ell m} = 0 . \qquad (6.78)$$

Here and in (6.76) set $d^{-2}_{\ell m} = d^{-1}_{\ell m} = \tilde{d}^{-2}_{\ell m} = \tilde{d}^{-1}_{\ell m} = 0$. The recursion relations (6.76 & 78) may in principle be employed to find the eigenvalue expansions started in Eq. (6.68) as power series in $a\omega$ up to any order desired.

In the special case $s = 0$, this is possible as then Eqs. (6.76 & 78) reduce to three-term recursion relations, which can be solved by the following standard method (cf. Morse & Feshbach 1963; Meixner & Schäfke 1954). The pair of eigenvalues $({}_0A_{\ell m}, a^2\omega^2)$ are solutions of a certain equation of continued fractions derived from the recursion relations. For small values of a they can be replaced by power series. Then one inserts ${}_0A_{\ell m} = \ell(\ell+1) + O(a^2\omega^2)$ into the inverted continuous fractions and obtains higher order terms by successive iteration. This method leads to the following low frequency expansion for the eigenvalues ${}_0A_{\ell m}$ (Bouwkamp 1950):

$$ {}_0A_{\ell m} = \ell(\ell+1) - \frac{a^2\omega^2}{2} \left[1 + \frac{(2m-1)(2m+1)}{(2\ell-1)(2\ell+3)} \right] +$$

$$+ \frac{a^4\omega^4}{2} \left[\frac{(\ell-m-1)(\ell-m)(\ell+m-1)(\ell+m)}{(2\ell-3)(2\ell-1)^3(2\ell+1)} - \frac{(\ell-m+1)(\ell-m+2)(\ell+m+1)(\ell+m+2)}{(2\ell+1)(2\ell+3)^3(2\ell+5)} \right] -$$

$$- a^6\omega^6(4m^2-1) \left[\frac{(\ell-m-1)(\ell-m)(\ell+m-1)(\ell+m)}{(2\ell-5)(2\ell-3)(2\ell-1)^5(2\ell+1)(2\ell+3)} - \frac{(\ell-m+1)(\ell-m+2)(\ell+m+1)(\ell+m+2)}{(2\ell-1)(2\ell+1)(2\ell+3)^5(2\ell+7)} \right]$$

$$+ O(a^8\omega^8) . \qquad (6.79)$$

However, for $s \neq 0$ a suitably extended procedure has to be used to solve the general five-term recursion relation (for details see Breuer & Ryan 1975).

6.13 High-Frequency Expansion of Spin-Weighted Spheroidal Harmonics

The spheroidal wave equation (6.25) is given by

$$\left[\frac{d}{dx}(1-x^2)\frac{d}{dx} + a^2\omega^2x^2 - 2sa\omega x - \frac{(m+sx)^2}{1-x^2} + {}_sA+s\right] {}_sS(-a\omega,x) = 0, \tag{6.80}$$

where $x = \cos\theta$. In this section solutions of (6.80) are found for $a^2\omega^2 \to \pm\infty$. Here the plus sign corresponds to real frequencies, the minus sign to imaginary frequencies. The method used is analogous to that given in Erdelýi (1955), Flammer (1957) and Meixner & Schäfke (1954). The leading terms in $a\omega$ for both eigenfunctions and eigenvalues will be derived. For higher order terms see Breuer & Ryan (1975). One may think of the real-frequency case as the one which is physically important once the stability of Kerr black holes is granted.

Large real frequencies $a^2\omega^2 \to +\infty$

The aim is to transform Eq. (6.80) in such a way that in the limit $a\omega \to \pm\infty$ it can be compared with a differential equation whose solutions are known. A suitable transformation is

$${}_sS(x) = (1-x^2)^{\frac{m+s}{2}} g(x), \qquad u = 2a\omega(1-x), \tag{6.81}$$

under which Eq. (6.80) becomes

$$4a\omega\left[ug''(u) + (m+s+1)g'(u) - \tfrac{1}{4}(u+2s)g(u)\right] - \left[u^2g''(u) + \right.$$
$$\left. + 2(m+s+1)g'(u) + \left\{ms\frac{1-u/2a\omega}{1-u/4a\omega} - \frac{u^2}{4} - su - {}_sA + m(m+1) - a^2\omega^2\right\} g(u)\right] = 0. \tag{6.82}$$

The transformation of variables in (6.81) changes the regular singular points $x = \pm 1$ to $u = 0, +\infty$ in the limit $a^2\omega^2 \to \infty$. Multiplying Eq. (6.82) by $1/4a\omega$ and formally setting $1/a\omega = 0$ reduces it to

$$ug''(u) + (m+s+1)g'(u) + \left[{}_sA^* - \frac{u}{4}\right] g(u) = 0 \; , \qquad (6.83)$$

where

$${}_sA^* = \frac{1}{4a\omega}\left[{}_sA + a^2\omega^2 - 2sa\omega - m^2 - 2sm\right].$$

Eq. (6.83) may be compared with an equation which can be derived from the one for generalized Laguerre-polynomials $L_p^{(n)}(u)$ by the transformation

$$v(u) = e^{-u/2} L_p^{(n)}(u) \, , \; L_p^{(n)}(u) = \frac{e^u u^{-n}}{p!} \frac{d^p}{dx^p}\left(e^{-u} u^{n+p}\right).$$

This equation is

$$uv''(u) + (n+1)v'(u) + \;\; p + \frac{n+1}{2} - \frac{u}{4} \;\; v(u) = 0. \qquad (6.84)$$

Clearly, the following identifications can be made by comparing (6.84) with (6.83):

$$n = m + s, \qquad {}_sA^* = p + \frac{m + s + 1}{2} \quad . \qquad (6.85)$$

The quantity p has yet to be determined. The function $g(u) = (1-x^2)^{-\frac{m+s}{2}} {}_sS$ is a bounded solution of Eq. (.83) for x in the interval $(-1,1)$. Therefore, $v(u)$ must be a bounded solution of Eq. (6.84). This is the case only when p is an integer. Then p equals the number of zeros of $L_p^{(m+s)}(u)$, and hence of $v(u)$, in the interval $(-1,1)$. In order to find this number p the following theorem is required:

<u>Theorem 6.1</u>

The number of zeros p of ${}_sS_{\ell m}(-a\omega,x)$ for x in the interval $(-1,1)$ and for real $a\omega$ is independent of $a\omega$ and equal to $(\ell-m-s)$.

Since ${}_sS_{\ell m}(-a\omega,x)$ is either an even or an odd function in x it must be of the form

$$y_p^{\pm}(x) = (1-x^2)^{\frac{m+s}{2}}\left[e^{-u/2} L_p^{(m+s)}(u) \pm e^{-u/2} L_p^{(m+s)}(u^*)\right] \qquad (6.86)$$

for $a\omega > 0$, $p = 0,1,2,\ldots$, where $u = 2a\omega(1-x)$, $u^* = 2a\omega(1+x)$. In order to make $y_p^{\pm}(x)$ an approximation of ${}_sS_{\ell m}$, their respective number of zeros has to be the same. Hence set

$$2p = \begin{cases} \ell - m - s & \text{for } \ell-m-s \text{ even,} \\ \ell - m - s - 1 & \text{for } \ell-m-s \text{ odd}. \end{cases} \tag{6.87}$$

Finally $y_p^{\pm}(x)$ has to be normalized appropriately in accordance with the normalization of ${}_sS_{\ell m}$ as given in (6.23). The result for the leading term in the eigenfunction expansion $a^2\omega^2 \to +\infty$ is given by

$$\begin{aligned} {}_sS_{\ell m}(-a\omega,x) = {}_sC_{\ell m}(1-x^2)^{\frac{m+s}{2}} \Big[& e^{-u/2} L_p^{(m+s)}(u) + \\ & +(-1)^{\ell-m-s} e^{-u^*/2} L_p^{(m+s)}(u^*) \Big] + O(1/a\omega), \end{aligned} \tag{6.88}$$

where p is defined in (6.87), $u = 2a\omega(1-x)$, $u^* = 2a\omega(1+x)$ and

$${}_sC_{\ell m} = (-1)^{m+s} (a\omega)^{\frac{m+s+1}{2}} \left[\frac{1}{2\ell+1} \frac{(\ell+m)!}{(\ell-m)!} \frac{p!}{(p+m+s)!} \right]^{1/2}.$$

Using (6.85), the leading term of the corresponding eigenvalue expansion is the following,

$${}_sA_{\ell m} = \begin{cases} 2(\ell+s+1)a\omega + m^2 - a^2\omega^2 + 2sm\omega + O(1), & \ell = m+s, m+s+2, \ldots \\ 2(\ell+s)a\omega + m^2 - a^2\omega^2 + 2sm\omega + O(1), & \ell = m+s+1, m+s+3, \ldots \end{cases} \tag{6.89}$$

Higher order terms can be obtained using expressions (6.88) as basis functions for an expansion in lower orders of $a\omega$.

Large imaginary frequencies $a^2\omega^2 \to -\infty$

When considering imaginary frequencies one puts $a\omega = ia\omega^*$ and lets $a^2\omega^{*2} \to +\infty$. Applying transformations

$${}_sS(x) = (1-x^2)^{\frac{m+s}{2}} g(x), \qquad u = \sqrt{2a\omega^*}\, x, \tag{6.90}$$

to Eq. (6.80) yields

$$2a\omega^* \left[g''(u) - \left(\frac{u^2}{4} + \frac{is}{\sqrt{2a\omega^*}}u\right)g(u)\right] + \\ + \left[{}_sA - m(m+1) - \frac{1}{2a\omega^*}\frac{2smu}{1+u/\sqrt{2a\omega^*}}\right]g(u) = 0 \,. \tag{6.91}$$

The regular singular points $x=\pm 1$ become $u=\pm\infty$ for $a\omega^*\to\infty$. If one divides Eq. (6.91) by $2a\omega^*$, then for large $a\omega^*$ this equation approaches

$$g''(u) + \left[{}_sA^* - \frac{u^2}{4}\right] g(u) = 0,$$

where (6.92)

$${}_sA^* = \frac{1}{2a\omega^*}\left[{}_sA - m(m+1)\right] \,.$$

Solutions of Eq. (6.85) are parabolic cylinder-functions $D_p(u)$ given by

$$D_p(u) = (-1)^p e^{u^2/4} \frac{d^p}{dx^p} e^{-u^2/2}$$

provided one identifies Eq. (6.92) with

$$D_p''(u) + \left[p + \frac{1}{2} - \frac{u^2}{4}\right] D_p(u) = 0.$$

Clearly, the leading term in the eigenvalue expansion yields ${}_sA^* = p+\frac{1}{2}$ or

$${}_sA_{\ell m} = (2p+1)a\omega^* - m(m+1) + O(1)\,. \tag{6.93}$$

The relation between p and (ℓ,m,s) is established by demanding that the function

$$y_p(x) = (1-x^2)^{\frac{m+s}{2}} D_p\left[\sqrt{2a\omega^*}\, x\right]$$

and the function ${}_sS_{\ell m}$ have the same number of zeros for x in the interval (-1,1). Thus by theorem 6.1 which also holds for $a\omega^*$ instead of $a\omega$, $p = \ell - m - s$. After normalization of the function $y_p(x)$ one obtains for the leading term in the eigenfunction expansion for large negative $a^{2\ 2}$ the expression

$$
{}_sS_{\ell m}(-a\omega, x) = {}_sC_{\ell m}(1-x^2)^{\frac{m+s}{2}} D_{\ell-m-s}\left[\sqrt{2a\omega^*}\,x\right] + O(1/a\omega^*), \tag{6.94}
$$

where

$$
{}_sC_{\ell m} = (-1)^{m+s}\left[\frac{a\omega^*}{\pi}\right]^{1/4}\left[\frac{2}{2\ell+1}\,\frac{(\ell+m)!}{(\ell-m)!}\,(\ell-m-s)!\right]^{1/2}.
$$

Using (6.94) as a basis for an expansion, lower order terms in $a\omega^*$ for ${}_sS_{\ell m}$ may be found and similarly for the eigenvalues.

VII. POLARIZATION

7.1 Introduction

Calculations done with scalar radiation as a model for higher spin waves suffer from the fact that they are unable to give information on the polarization properties of radiation. Before studying explicitly the polarization of GSR, a general formalism for treating polarized radiation in General Relativity is developed, and some of its features, such as the equation of radiative transfer, are explored.

In many situations encountered in relativistic astrophysics one considers problems in which the transfer of electromagnetic radiation plays an essential role, e.g. in discussing the propagation of the cosmic background radiation in cosmology, the internal structure of supermassive stars, or the radiation emitted near the surface of a neutron star (or a black hole). Lindquist (1966), Ellis (1971), and others have investigated the transfer of unpolarized electromagnetic radiation in GR, by using either kinetic theory or geometrical optics.

The basis of the treatment of polarized radiation in this chapter is the definition of the so-called Stokes Parameters (SP) for General Relativity, which form a set of four (observer dependent) scalars describing the polarization properties of a wave. In Sec. 3 and 4 an invariant definition of polarization tensors for high frequency electromagnetic and gravitational radiation is given. The SP are then defined by projecting this tensor onto the screen-plane orthogonal to the wave's propagation direction and the observer's velocity. A polarization matrix ("coherence matrix") analogous to the density matrix of statistical mechanics is introduced, whose trace yields the energy flux across the screen-plane. From that, with the help of the propagation equations for the respective field quantities, the equation of radiative transfer is derived for polarized electromagnetic and gravitational radiation in GR in terms of the SP (Sec. 5), for the latter case both in vacuo and in the presence of a dissipative fluid. In Sec. 6, the SP are generalized to waves occurring in the most general class of metric theories of gravitation. Since there are six independent states of polarization present, they are described either by a Hermitian 6x6 polarization matrix, or, equivalently, by 36 Stokes Parameters.

Before plunging into the formalism a review of the standard treatment of polarization in ordinary electromagnetic theory is given.

7.2 Stokes Parameters, Poincaré Sphere, Jones Vector

In flat space electromagnetism several methods have been developed to obtain a unique representation of a (partially) polarized wave. Sir George Stokes (1852) introduced four observer-dependent scalars in his paper, "On the Composition and Resolution of Streams of Polarized Light from Different Sources." Those parameters, which are the observables of the radiation field, were later used to describe polarization in a three-dimensional space - the so-called Poincaré sphere - by Poincaré (1892). They were rediscovered and finally called Stokes Parameters by Chandrasekhar (1950). More extensive tratments of electromagnetic SP can be found in the books by Jauch & Rohrlich (1955), O'Neill (1963), Shurcliff & Ballard (1964), and Clarke & Graininger (1971); for application to radio astronomy, see Kraus (1966).

Jones (1941) introduced a method of treating polarization in terms of the (complex) amplitudes of the radiation field. Consider a totally polarized (monocromatic) wave travelling in the x^3-direction. The Jones vector $\underline{E}$ is then given by the two complex, transverse wave amplitudes

$$\underline{E} = \begin{pmatrix} E_1 \\ E_2 \end{pmatrix} . \tag{7.1}$$

Changes in polarization (e.g., due to the action of an instrument on the wave) are represented as complex linear operators (complex 2x2 matrices) acting on $\underline{E}$. This way of describing radiation by complex two-vectors can, however, only be applied to the totally polarized part of the radiation; it is not suitable for dealing with the superposition of partially or unpolarized waves. Also, the complex amplitudes are, of course, not the observables of the field.

In order to handle partial polarization it seems more appropriate to use the so-called coherence-matrix formalism, which is constructed in analogy with the density matrix methods of quantum statistical mechanics. The coherence matrix J is defined by

$$J = (\underline{E} \times \underline{E}^{+})_{av} = \begin{bmatrix} (E_1\bar{E}_1)_{av} & (E_1\bar{E}_2)_{av} \\ (\bar{E}_1E_2)_{av} & (E_2\bar{E}_2)_{av} \end{bmatrix} \equiv \begin{bmatrix} J_{11} & J_{12} \\ J_{21} & J_{22} \end{bmatrix} , \quad (7.2)$$

where $\underline{E}^{+} = \underline{\bar{E}}^{T} = (\bar{E}_1, \bar{E}_2)$. The bracket $(\ldots)_{av}$ denotes time averaging (ensemble average). For unpolarized radiation, no direction of the electric field vector is preferred and on the average there will be no correlation between the two components. Hence $(E_1\bar{E}_2)_{av} = (E_1)_{av}(\bar{E}_2)_{av} = 0$ and J will be diagonal for unpolarized radiation, i.e.

$$J^{u} = \frac{1}{2} I \begin{pmatrix} 1 & 0 \\ 0 & 1 \end{pmatrix}, \quad \text{where} \quad I = (E_1\bar{E}_1)_{av} + (E_2\bar{E}_2)_{av} \quad .$$

On the other hand, radiation is totally polarized when $\det(J) = 0$, i.e. when $J_{ij} = E_i \cdot E_j$ (without averaging); the degree of polarization d is then defined by $d \equiv J^{p}/(J^{u} + J^{p}) \equiv J^{p}/I^{tot}$, where J^{p} is the intensity of the polarized part of the radiation. In general (partial polarization), the coherence matrix (7.2) can be decomposed uniquely into a totally polarized and a totally unpolarized part, namely

$$J_{ij} = \frac{1}{2} (I^{tot} - J^{p}) \delta_{ij} + J^{p} E_i\bar{E}_j \quad . \quad (7.3)$$

Thus any partially polarized quasi-monochromatic wave can be thought of as an incoherent superposition of a completely unpolarized wave and a fully polarized monochromatic wave.

The coherence matrix has another decomposition as an expansion in terms of the Pauli matrices. Using the representation (O'Neill 1973)

$$\sigma_o = \begin{pmatrix} 1 & 0 \\ 0 & 1 \end{pmatrix}, \sigma_1 = \begin{pmatrix} 1 & 0 \\ 0 & -1 \end{pmatrix}, \sigma_2 = \begin{pmatrix} 0 & 1 \\ 1 & 0 \end{pmatrix}, \sigma_3 = \begin{pmatrix} 0 & i \\ -i & 0 \end{pmatrix} \quad (7.4)$$

one can write the expansion as

$$J = \frac{1}{2} \sum_{A=0} S_A \sigma_A = \frac{1}{2}(S_o \sigma_o + S_1 \sigma_1 + S_2 \sigma_2 + S_3 \sigma_3) . \quad (7.5)$$

To solve for the expansion coefficients S_A, multiply (7.5) by σ_B on both sides, take the trace, and use the relation $tr(\sigma_A \sigma_B) = 2\delta_{AB}$ to get

$$S_A = tr(J \cdot \sigma_A) . \quad (7.6)$$

With the aid of Eq. (7.2) this may be written more explicitly as

$$\begin{pmatrix} S_o \\ S_1 \\ S_2 \\ S_3 \end{pmatrix} = \begin{pmatrix} 1 & 0 & 0 & 1 \\ 1 & 0 & 0 & -1 \\ 0 & 1 & 1 & 0 \\ 0 & -i & i & 0 \end{pmatrix} \begin{pmatrix} J_{11} \\ J_{12} \\ J_{21} \\ J_{22} \end{pmatrix} . \quad (7.7)$$

These four scalars are, by definition, the <u>Stokes</u> <u>Parameters</u> of the electromagnetic field. Inserting (7.7) into (7.5), the coherence matrix becomes

$$J = \frac{1}{2} \begin{pmatrix} S_o+S_1 & S_2+iS_3 \\ S_2-iS_3 & S_o-S_1 \end{pmatrix} . \quad (7.8)$$

Solution (7.6) for the SP suggests an obvious interpretation: the polarization states form a two-dimensional complex Hilbert space with the coherence matrix as a representation, in which the four Pauli matrices are the independent states of polarization. The SP then are the expectation values of these polarization states. They provide a different parametrization of the coherence matrix.

The total intensity is $I^{tot} = tr(J\,\sigma_o) = J_{11} + J_{22} = S_o$. For the superposition of incoherent beams of light, it suffices to add the SP

Table VI. Jones vector, coherence matrix and Stokes "vector" for different polarization states. (from O'Neill 1963)

State of polarization	$\boldsymbol{\varepsilon}$	$\mathbf{J}$	$\mathbf{S}$
Plane of polarization in the x-direction	$\begin{bmatrix}1\\0\end{bmatrix}$	$\begin{bmatrix}1&0\\0&0\end{bmatrix}$	$\begin{bmatrix}1\\1\\0\\0\end{bmatrix}$
Plane of polarization in the y-direction	$\begin{bmatrix}0\\1\end{bmatrix}$	$\begin{bmatrix}0&0\\0&1\end{bmatrix}$	$\begin{bmatrix}1\\-1\\0\\0\end{bmatrix}$
Plane of polarization at 45° to the x-axis	$\begin{bmatrix}1\\1\end{bmatrix}$	$\begin{bmatrix}1&1\\1&1\end{bmatrix}$	$\begin{bmatrix}1\\0\\1\\0\end{bmatrix}$
Plane of polarization at 135° to the x-axis	$\begin{bmatrix}1\\-1\end{bmatrix}$	$\begin{bmatrix}1&-1\\-1&1\end{bmatrix}$	$\begin{bmatrix}1\\0\\-1\\0\end{bmatrix}$
Right circular polarization	$\begin{bmatrix}1\\-i\end{bmatrix}$	$\begin{bmatrix}1&i\\-i&1\end{bmatrix}$	$\begin{bmatrix}1\\0\\0\\1\end{bmatrix}$
Left circular polarization	$\begin{bmatrix}1\\i\end{bmatrix}$	$\begin{bmatrix}1&-i\\i&+1\end{bmatrix}$	$\begin{bmatrix}1\\0\\0\\-1\end{bmatrix}$

of the single beams. Here, the method is complementary to the Jones vector approach. For coherent superposition of light beams, the SP method in general gives wrong answers; this case has to be dealt with using the Jones vectors.

In terms of the SP the degree of polarization d is

$$d = (S_1^2 + S_2^2 + S_3^2)^{1/2}/S_o. \tag{7.9}$$

As generally, $0 \leq d \leq 1$, Eq. (7.9) gives the relation

$$S_o^2 \geq S_1^2 + S_2^2 + S_3^2 . \tag{7.10}$$

In the fully polarized case, equality holds (d=1) in Eq. (7.10). Eq. (7.10) is then the equation of a sphere in the Stokes subspace

S_1, S_2, S_3 with the radius S_o. This sphere is called the Poincaré sphere. Usually, S_o is normalized to unity so that the Poincaré sphere has unit radius (see Fig. 10). For some special cases of purely monochromatic waves in definite states of polarization, the values of Jones vector, coherence matrix, and Stokes Parameters are listed in Table VI.

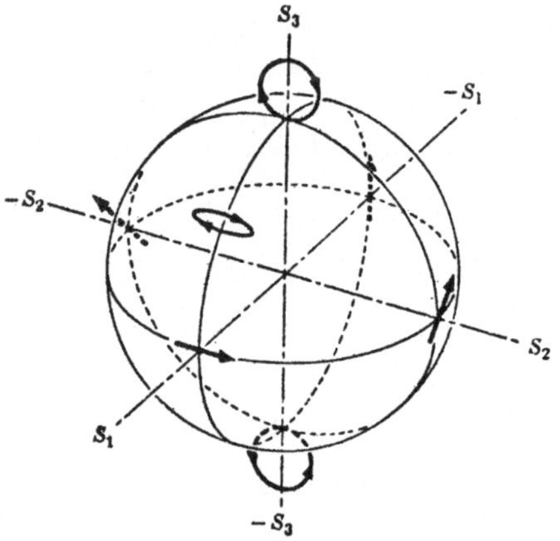

Figure 10. Poincaré sphere (normalized Stokes Parameters). For partially polarized waves the radius of the sphere is S_o (degree of polarization); therefore the radius tends to zero for unpolarized waves. The north and south poles represent left and right circular polarization, respectively$^+$. Every point on the equator stands for a different linearly polarized state (S_1 and S_2). Points between the equator and the north pole represent left-handed elliptical polarization (right-handed in the southern hemisphere). S_3 adds circular polarization components to the radiation. The action of polarizers is visualized by definite position changes of the polarization "vector" on the Poincaré sphere. (from O'Neill 1963)

$^+$ A wave is right circularly polarized, when the electric field vector is rotating counter-clockwise with the wave approaching or clockwise with the wave receding.

7.3 Electromagnetic Stokes Parameters

The generalization of Stokes Parameters to curved spacetime for electromagnetic and gravitational fields perturbing an arbitrary vacuum background gravitational field was given by Anile & Breuer (1974) in a coordinate dependent formalism. Here, a brief account of the electromagnetic and gravitational definitions is given in an invariant manner, thus exhibiting more clearly certain invariance properties of Stokes parameters[+].

Consider first a locally plane (null) electromagnetic wave

$$F_{ab} = \mathrm{Re}\left\{ 2k_{[a}m_{b]}\, e^{iS} \right\}, \tag{7.11}$$

where S is the phase, $k_a \equiv S_{,a}$ is a real null vector, m_a is complex and orthogonal to k_a, and

$$\begin{aligned} k_a k^a &= k_a m^a = Dk_a = 0, \\ Dm_a &+ \theta m_a = 0 . \end{aligned} \tag{7.12}$$

Here, D is the covariant derivative along k_a and $\theta \equiv \frac{1}{2}\nabla^a k_a$ is the expansion of the light rays. Because of Eq. (7.12) the quantities m_a and S are unique only up to a "gauge" transformation; they belong respectively to the equivalence classes

$$\left\{ \tilde{m}_a : \tilde{m}_a = e^{i\delta} m_a + \lambda k_a \right\}, \quad \left\{ \tilde{S} : \tilde{S} = S - \delta \right\}, \tag{7.13}$$

where $\lambda, \delta \in \mathbb{R}$. At a given point x of spacetime and for a fixed vector field k_a, the complex vectors m^a subject to $k^a m_a = 0$ form a two-dimensional complex vector space. In this vector space a positive semi-definite Hermitian product is defined by

$$\begin{aligned} \langle m_1, m_2 \rangle &= \bar{m}_1{}^a m_{2a}, \\ \text{with} \quad \langle m, m \rangle &= 0 \Leftrightarrow m^a = \lambda k^a. \end{aligned} \tag{7.14}$$

[+] We thank Prof. J. Ehlers for improving the formulation of this section.

This shows, that the gauge-equivalence classes of vectors m at a point x as defined in (7.13) form a two-dimensional Hilbert space $\mathcal{H}_x^k$ with respect to the product $\langle\, , \rangle$.

$\mathcal{H}_x^k$ is defined in a coordinate and observer-independent way and is intrinsically given by the ray congruence $\{k^a\}$. Given an observer with four-velocity u^a at x, there exists in each m-equivalence class exactly one member satisfying $m^a u_a = 0$, where one has chosen a fixed phase such that $\delta = 0$. An element of $\mathcal{H}_x^k$ with respect to some observer can therefore be represented exactly by one vector m^a with $m^a u_a = 0$. In fact, with these representatives chosen in this way, $\mathcal{H}_x^k$ is the space of Jones vectors, but for all observers simultaneously.

Given an ensemble of vectors m^a or "waves" F_{ab}, with fixed phase S and null direction k^a, form the following product averaged over the ensemble,

$$L_{ab} = (m_a \bar{m}_b)_{av} \quad . \tag{7.15}$$

Clearly, L_{ab} defines a coordinate- and observer-independent Hermitian operator $\mathbb{L}(x)$ in $\mathcal{H}_x^k$. The quantities $k^a(x)$ and $\mathbb{L}(x)$ completely describe the wave ensemble with respect to their energy and polarization properties. The ensemble average of the energy-momentum tensor of the wave is

$$(T_{ab})_{av} = \frac{1}{2}\,(\mathrm{Tr}\ \mathbb{L})\, k_a k_b , \tag{7.16}$$

where Tr is the trace operation in $\mathcal{H}_x^k$. To obtain an orthonormal basis $(t,\bar{t})$ in $\mathcal{H}_x^k$, one takes any complex vector t^a satisfying $\langle t,t\rangle = \bar{t}^a t_a = 1$ together with its complex conjugate $\bar{t}^a$.

The Stokes parameters with respect to such a basis $(t,\bar{t})$ and with respect to an observer with four-velocity u^a satisfying $u^a t_a = 0$ can be defined by

$$\begin{aligned} S_0 &= \omega^2\ \mathrm{Tr}\ \mathbb{L} , \\ S_1 &= \omega^2 \left[\langle t, \mathbb{L}\, t\rangle - \langle \bar{t}, \mathbb{L}\, \bar{t}\rangle\right], \\ S_2 + i\, S_3 &= 2\,\omega^2 \langle \bar{t}, \mathbb{L}\, t\rangle , \end{aligned} \tag{7.17}$$

where $\omega = |u_a k^a|$. Eqs. (7.17) imply that the quantities $F_A = S_A/\omega^2$ $(A = 0,\dots,3)$ are independent of u^a and depend only on the basis $(t,\bar{t})$. A rotation of the basis by an angle δ, i.e. $(t,\bar{t}) \to (e^{i\delta}t, e^{-i\delta}\bar{t})$, leaves F_o, F_1 invariant; $F_2 + iF_3$ picks up a factor $\exp(2i\delta)$.

With the aid of the normalized density matrix $\mathbb{W} = (\mathrm{Tr}\,\mathbb{L})^{-1}\,\mathbb{L}$, the normalized Stokes parameters are defined by $s_A = S_a/S_o$, or, more explicitly, by

$$s_o = 1, \qquad s_1 = \langle t, \mathbb{W}t\rangle - \langle \bar{t}, \mathbb{W}\bar{t}\rangle,$$
$$s_2 + is_3 = 2\langle \bar{t}, \mathbb{W}t\rangle. \tag{7.18}$$

The luminosity distance r is defined by the relation $\frac{1}{r}\, r_{,a}k^a = 0$ (cf. Pirani 1964). Then the operator $r^2\,\mathbb{L}$ has constant matrix elements along the ray and with respect to a parallelly propagated basis $(t,\bar{t})$. Hence $r^2 S_A/\omega^2$ is constant along the rays.

For a <u>general</u> non-null radiation field, i.e. an ensemble of waves (or photons) with different k^a-fields and corresponding $\mathbb{L}$'s, introduce the invariant measure $\pi_o \equiv d^3k/k^o$ on the future light cone. To each "small area" K on the null-cone at x containing the end-points of the null vectors k^a there corresponds

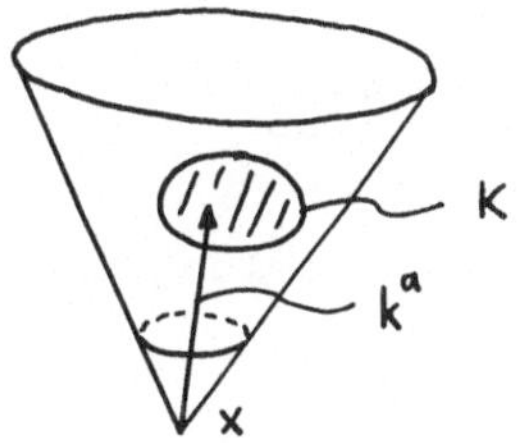

an operator $\mathbb{L}(k)$ as defined after Eq. (7.15); put

$$\lim_{K \to k^a} \frac{\mathbb{L}(K)}{\pi_o(k)} = \mathbb{F}(x^a, k^a).$$

The domain of definition of this function $\mathbb{F}(x^a,k^a)$ is the bundle of nullcones $\{(x^a,k^a)\}$, and its range consists again of Hermitian operators in $\mathcal{H}_x^k$. This invariantly defined <u>distribution matrix</u> completely describes the radiation field. The ensemble-averaged energy-momentum tensor at x is

$$(T_{ab}(x))_{av} = \frac{1}{2} \int\limits_{\begin{pmatrix}\text{light cones}\\ \text{in } x\end{pmatrix}} \pi_o (\mathrm{tr}\, \mathbb{F}) k_a k_b \equiv \int \pi_o\, f(x,k) k_a k_b . \tag{7.19}$$

The function $f(x,k) = \frac{1}{2} \mathrm{Tr}\, \mathbb{F}$ is the usual relativistic distribution function for photons (irrespective of their polarization). The definition of the SP for such a mixture of photons is identical to Eqs. (7.17) with $\mathbb{F}$ replacing $\mathbb{L}$, provided that one deals with SP densities.

7.4 Gravitational Stokes Parameters

Consider first a locally plane null gravitational wave. Then the Weyl tensor of the perturbation is given by

$$\delta C_{abcd} = -2\,\mathrm{Re}\left\{ k_{[a} m_{b]} {}_{[c} k_{d]}\, e^{iS} \right\} , \tag{7.20}$$

where S is a real phase factor, m_{ab} is a complex bivector and

$$\begin{aligned} k_a &= S_{,a} , \quad m_{ab} = - m_{ba} , \quad m_{ab} k^b = 0 , \\ m_a{}^a &= 0 , \quad D m_{ab} + \theta m_{ab} = 0. \end{aligned} \tag{7.21}$$

The phase S and the bivector m_{ab} belong to the equivalence classes

$$\begin{aligned} &\left\{ \tilde{m}_{ab} : \tilde{m}_{ab} = e^{i\delta} m_{ab} + \lambda_{(a} k_{b)} , \quad \lambda_a \in \mathbb{R} , \lambda_a k^a = 0 \right\} , \\ &\left\{ \tilde{S} : \tilde{S} = S - \delta , \delta \in \mathbb{R} \right\} , \end{aligned} \tag{7.22}$$

respectively. The Hilbert space $\mathcal{H}_x^k$ at the spacetime point x for fixed k is the vector space of m_{ab}-equivalence classes with the positive semi-definite Hermitian inner product

$$\begin{aligned} \langle m_1, m_2 \rangle &= \frac{1}{2} \bar{m}_1{}^{ab} m_{2\,ab} , \\ \text{with} \quad \langle m, m \rangle &= 0 \Leftrightarrow m_{ab} = \lambda_{(a} k_{b)} . \end{aligned} \tag{7.23}$$

As in the electromagnetic case, in each equivalence class a specific m_{ab} can be found by a gauge transformation which satisfies $m_{ab}u^b = 0$. In the case of an ensemble of waves, introduce the polarization tensor

$$L_{abcd} = (m_{ab}\,\bar{m}_{cd})_{av} \tag{7.24}$$

which defines in a coordinate- and observer-independent way a Hermitian operator $\mathbb{L}(x)$ in $\mathcal{H}^k_x$. The Isaacson energy-momentum tensor of the ensemble is

$$(T_{ab})_{av} = \frac{1}{32\pi}\,(\mathrm{Tr}\,\mathbb{L})\,k_a k_b\ . \tag{7.25}$$

A complex vector $t^a(\bar{t}^a t_a = 1)$ defines an orthonormal basis (e_R, e_L) in $\mathcal{H}^k_x$ via the unit polarization tensors

$$\underset{=}{e}_R = \sqrt{2}\,t \otimes t, \qquad \underset{=}{e}_L = \sqrt{2}\,\bar{t} \otimes \bar{t}\ . \tag{7.26}$$

These tensors represent right- and left-handed circularly polarized waves. The states of linear polarization "+" and "×" are given by the unit tensor (MTW 1973)

$$\underset{=}{e}_+ = \frac{1}{\sqrt{2}}\,(\underset{=}{e}_R + \underset{=}{e}_L), \qquad \underset{=}{e}_x = \frac{1}{\sqrt{2}}\,(\underset{=}{e}_R - \underset{=}{e}_L)\ ,$$

respectively. The gravitational Stokes parameters are defined by

$$\begin{aligned} S_o &= \omega^2\,\mathrm{Tr}\,\mathbb{L}, \\ S_1 &= \omega^2\,(\langle \underset{=}{e}_R, \mathbb{L}\,\underset{=}{e}_R\rangle - \langle \underset{=}{e}_L, \mathbb{L}\,\underset{=}{e}_L\rangle)\ , \\ S_2 + i\,S_3 &= 2\,\omega^2 \langle \underset{=}{e}_L, \mathbb{L}\,\underset{=}{e}_R\rangle . \end{aligned} \tag{7.27}$$

Again, the quantities $F_A = 2S_A/\omega^2$ are independent of the observer's four-velocity u^a. A rotation $t \rightarrow e^{i\delta}t$ changes the basis according to

$$\underset{=}{e}_R \rightarrow e^{2i\delta}\,\underset{=}{e}_R\ , \qquad \underset{=}{e}_L = e^{-2i\delta}\,\underset{=}{e}_L .$$

Hence S_o, S_1 are invariant under such a rotation, while $S_2 + iS_3$ changes by a factor $\exp(4i\delta)$. The normalized density matrix is again $\mathbb{W} = (\mathrm{Tr}\,\mathbb{L})^{-1}$ and the normalized gravitational Stokes parameters are given by $s_A = S_A/S_o$, i.e. by

$$s_o = 1, \qquad s_1 = \langle \underset{=}{e}_R, \mathbb{W}\, \underset{=}{e}_R \rangle - \langle \underset{=}{e}_L, \mathbb{W}\, \underset{=}{e}_L \rangle ,$$
$$s_2 + i\, s_3 = 2 \langle \underset{=}{e}_L, \mathbb{W}\, \underset{=}{e}_L \rangle . \tag{7.28}$$

The matrix elements of $r^2\,\mathbb{L}$ are constant when propagated along the ray; hence so is $2r^2 s_A/\omega^2$.

For a <u>general</u> non-null radiation field one has, as in case of photons, the <u>distribution operator</u> $\mathbb{F}(x,k)$ with

$$(T_{ab}(x))_{av} = \frac{1}{32\pi} \int\limits_{\begin{pmatrix}\text{light cones}\\ \text{in } x\end{pmatrix}} (\mathrm{Tr}\,\mathbb{F}) k_a k_b\, \pi_o \equiv \int\limits_{\begin{pmatrix}\text{light cones}\\ \text{in } x\end{pmatrix}} f(x,k) k_a k_b\, \pi_o . \tag{7.29}$$

The function $f(x,k) = \frac{1}{32\pi}\,\mathrm{Tr}\,\mathbb{F}$ is the distribution function for gravitons independent of polarization. The definition for the SP in this general case is the same as in (7.27) provided $\mathbb{L}$ is replaced by $\mathbb{F}$.

So far the SP have been defined in an invariant manner, but now they shall be expressed with respect to a tetrad associated with an observer. Let the observer's four-velocity be u^a $(u^a u_a = -1)$ and choose a tetrad $\{e_{\hat{a}}{}^b\}$ such that the observer sees the wave travelling into the + z-direction. Thus,

$$e_{\hat{0}}{}^a = u^a , \qquad e_{\hat{3}}{}^a = (-u_b k^b)^{-1} \left[k^a + (u_b k^b)\, u^a \right] . \tag{7.30}$$

To evaluate the density matrix in the frame (7.30), define

$$J_{abcd} = \omega^2\, L_{abcd} = (E_{ab}\, \bar{E}_{cd})_{av} , \tag{7.31}$$

where $E_{ab} = \delta C_{abcd}\, u^c u^d$. The quantity J_{abcd} has the following properties, which are obvious from relations (7.21):

$$J_{abcd} = J_{(ab)(cd)}, \quad J_{abcd} = \overline{J}_{cdab} ,$$
$$J_{a\ cd}^{\ a} = J_{abc}^{\ \ \ \ c} = J_{abcd}k^d = 0. \tag{7.32a}$$

In addition, a gauge is chosen so that

$$J_{abcd}\ u^b = J_{abcd}\ u^d = 0. \tag{7.32b}$$

The elements of the density matrix are then given by

$$\mathbb{J} = \omega^2\ \mathbb{L} = \omega^2 \begin{pmatrix} L_{\hat{1}\hat{1}\hat{1}\hat{1}} & L_{\hat{1}\hat{1}\hat{1}\hat{2}} \\ \overline{L}_{\hat{1}\hat{1}\hat{1}\hat{2}} & L_{\hat{1}\hat{2}\hat{1}\hat{2}} \end{pmatrix}, \tag{7.33}$$

where $L_{\hat{\alpha}\hat{\beta}\hat{\gamma}\hat{\delta}} = L_{abcd}\ e_{\hat{\alpha}}^{\ a}\ e_{\hat{\beta}}^{\ b}\ e_{\hat{\gamma}}^{\ c}\ e_{\hat{\delta}}^{\ d}$ $(\alpha,\ldots,\delta = 1,2)$. Using Eq. (7.33) the Stokes parameters can now be written as

$$S_o = \omega^2\ \mathrm{Tr}\ \mathbb{L} = J_{\hat{1}\hat{1}\hat{1}\hat{1}} + J_{\hat{1}\hat{1}\hat{1}\hat{2}} ,$$
$$S_1 = J_{\hat{1}\hat{1}\hat{1}\hat{1}} - J_{\hat{1}\hat{1}\hat{1}\hat{2}} ,$$
$$S_2 = J_{\hat{1}\hat{1}\hat{1}\hat{2}} + J_{\hat{1}\hat{2}\hat{1}\hat{1}} , \tag{7.34}$$
$$S_3 = i(J_{\hat{1}\hat{1}\hat{1}\hat{2}} - J_{\hat{1}\hat{2}\hat{1}\hat{1}}).$$

For the normalized Stokes parameters $s_A = S_A/S_o$ relations (7.9 & 10) hold as they did for Electromagnetism in flat and curving spacetime. Eqs. (7.3-8) are equally valid when $\mathbb{J}$ and S_A are given by (7.33,34). Further, the following definitions are formally identical for EM and Gravitation. Degrees of linear and circular polarization are defined by

$$d_L = (s_1^2 + s_2^2)^{1/2} , \quad d_C = s_3 . \tag{7.35}$$

From the obvious properties of electromagnetic and gravitational SP under Lorentz transformations (see also discussions after Eqs. (7.17 & 27)) one may conclude the following result: the degrees of linear and circular polarization are invariant under Lorentz transformations for both electromagnetic and gravitational (high frequency) waves.

For Electromagnetism in flat spacetime this fact was first noted by Cocke & Holm (1972).

The inclination of the polarization ellipse defines some angle τ, namely the angle between its major axis and the positive x-axis of some fixed x-y coordinate axes that may be chosen arbitrarily within the plane orthogonal to the three-dimensional propagation direction $e^b_{\hat{3}}$ of the wave.

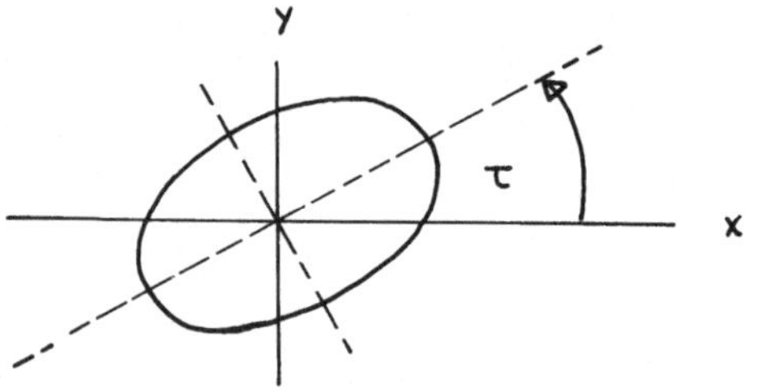

This angle is called the <u>tilt angle of the polarization ellipse</u> and is related to the Stokes parameters by

$$\tau = -\frac{i}{4}\,\ell n\,\frac{S_1^2\;S_2^2 + 2iS_1S_2}{S_1^2 + S_2^2}\,, \quad 0 \leq \tau < \pi . \tag{7.36}$$

The direction of the axes of the polarization ellipse coincide with the principal directions of the symmetric part of the matrix (7.33).

7.5 The Equations of Radiative Transport for Polarized Radiation

In order to find equations for radiative transfer one first has to know the propagation law for the corresponding polarization tensors appearing in the definition of the SP. It may be shown from (7.15 & 17) that in an arbitrary empty spacetime the polarization tensor L_{ab} satisfies the propagation law

$$DL_{ab} + 2\,\theta\,L_{ab} = 0. \tag{7.37}$$

Consider an example from cosmological context where one wishes to describe the observation of polarized radiation from a distant source.

In a given cosmological model the tetrad $\{e_{\hat{a}}\}$ at the emission point E is determined by inserting the 4-velocity of the emitter and the vector k^a into Eqs. (7.30). Then this tetrad is propagated parallelly to the observer O where a Lorentz transformation is made to the frame $\{e_{\hat{a}}\}$ with the aid of the 4-velocity of the observer and the vector k^a at O.

Transfer equation for polarized electromagnetic radiation

In order to proceed let us work in a cosmological framework and assume a fluid approximation (Ellis 1971). Let the field $u^a(x)$ be the 4-velocity of the fluid $(u^a u_a = -1)$. The field lines of $u^a(x)$ define the trajectories of the so-called fundamental observers. Select an arbitrary world line C among these field lines and a point $E \in C \cap G$, where G is a null geodesic. At E choose a tetrad $\{e_{\hat{a}}\}$ in the manner described above. Then parallelly propagate the frame along G according to $De_{\hat{a}}{}^b = 0$. It is easy to see that

$$DF_A + 2\theta F_A = 0 \quad \text{along } G. \tag{7.38}$$

Consider a bundle of neighbouring null geodesics around G with cross section dS. Obviously,

$$D(dS) + 2\theta dS = 0 \text{ along } G.$$

Hence

$$F_A dS = \text{const} \qquad \text{along } G,$$

or

$$F_A dS\Big|_{\text{emitter}} = F_A dS\Big|_{\text{observer}}. \tag{7.39}$$

The redshift z is defined by

$$1 + z = \frac{(k_a u^a)_{\text{emitter}}}{(k_a u^a)_{\text{observer}}}. \tag{7.40}$$

Combining $S_A = \frac{1}{2}(k_a u^a)^2 F_A$ with Eq. (7.39) shows that

$$S_A = \frac{K_A}{(1+z)^2\, dS}, \tag{7.41}$$

where K_A are four constants along the ray G. What is usually measured for extended sources is the flux per unit solid angle in some frequency interval $(\nu, \nu+d\nu)$; the definition of the SP is extended in standard manner and the $\hat{S}_A(\nu)$ are then defined as the SP per unit solid angle and unit frequency. In the unpolarized case these $\hat{S}_A(\nu)$ are called radiation intensities. Following closely the method of Ellis (1971) one can show, that in the case of radiation not interacting with matter, one gets

$$\frac{d}{dv}\hat{S}_A(\nu) = \frac{3}{1+z}\hat{S}_A(\nu)\frac{dz}{dv} . \qquad (7.42a)$$

Here, $\nu = \nu_E/(1+z)$, z is the redshift of E with respect to O, and v is an affine parameter along the geodesic connecting the emission point E with the observer O, i.e. E is a fundamental observer at an affine distance v from O. It is more useful to express Eq. (7.42a) with the help of quantities $\hat{F}_A = 2\hat{S}_A/\omega^2$ which are observer-independent scalars and which are constants along the rays G (cf. Ehlers 1973). In the monochromatic case this is equivalent to the constancy of $2r^2 S_A/\omega^2$ (see the remark following Eq. (7.18)). With the aid of (7.40 & 41) and the relation

$$\left[\hat{S}_A\right]_O = (1+z)^{-3}\left[\hat{S}_A\right]_E$$

Eq. (7.42a) can be rewritten as

$$\frac{d}{dv}\hat{F}_A = 0 \qquad \text{or} \qquad \frac{d}{dv}\mathbb{F} = 0. \qquad (7.42b)$$

Here it is understood that the components of the distribution matrix $\mathbb{F}$ with respect to the parallelly propagated frame are substituted into $d\,\mathbb{F}/dv = 0$. Considering absorption, emission, and scattering of radiation, one obtains the following equation:

$$\frac{d}{dv}\hat{F}_A = -\kappa\,\hat{S}_A(\nu) - \hat{E}_A(\nu) + S_A^+(\nu) . \qquad (7.43)$$

In this equation the $\hat{E}_A(\nu)$ represents the SP of the radiation emitted per unit volume, frequency, and solid angle during the interaction; the $\hat{S}_A^{\,+}(\nu)$ are the SP per unit frequency of the radiation scattered into the solid angle under consideration. The quantity κ is the absorption coefficient. Detailed expressions for the right-hand side in Eq. (7.43) have to be considered for each individual case. For various types of moving scatterers Acquista & Anderson (1974) have worked out $\hat{S}^{+}_{A}(\nu)$.

Gravitational Radiative Transfer

Let us first consider the case in which gravitational waves propagate in a vacuum and then the case in which they propagate through a dissipative fluid. From Eq. (7.24) one obtains the vacuum propagation law for the polarization tensor L_{abcd}

$$DL_{abcd} + 2\theta L_{abcd} = 0. \tag{7.44}$$

Proceeding as in the electromagnetic case, the vacuum equation of radiative transfer is found to be

$$\frac{d\hat{F}_A}{dv} = 0 \ , \tag{7.45}$$

where $\hat{F}_A = \hat{S}_A/\omega^3$. The other symbols in Eq. (7.45) have the same meaning as in (7.42) but the obvious generalizations for the SP are used.

In the case of a gravitational wave interacting with a dissipative fluid, Madore (1973) has shown that

$$2Dh_{ab} - \theta h_{ab} = -2\eta(u_c k^c h_{ab} - \chi_{[a}k_{b]}), \tag{7.46}$$

where η is the shear viscosity of the fluid and

$$\chi_a = h_{ab}u^b - \frac{k_a}{2u_b k^b} h_{cd}u^c u^d.$$

Hence the propagation law (7.44) becomes

$$DL_{abcd} - 2(\theta - \eta\omega)L_{abcd} = 0. \tag{7.47}$$

Propagating the tetrad $e_{\hat{a}}{}^{b}$ parallelly along the null geodesic k_a, it follows with the aid of (7.34) that

$$DS_A - 2(\theta - \eta\omega)S_A = 0 . \tag{7.48}$$

This equation makes it again obvious that the normalized Stokes parameters are constant along the rays.

7.6 Stokes Parameters for the Most General Theory of Gravity

In this section the generalization of the SP formalism is given for the most general class of metric theories of gravity where six independent states of polarization are present. For a discussion of this topic with respect to gravitational-wave experiments, see Eardley et al. (1973). As pointed out in that paper, the Riemann tensor of the most general wave can be decomposed into a sum over six modes of polarization. The "electric" components R_{iojo} $(i,j = 1,2,3)$ that govern the driving forces in a wave-detector, are expanded as

$$R_{iojo} = \sum_{s=1}^{6} a(s)\, e_{iojo}(s) , \tag{7.49}$$

where the $a(s)$ are the complex amplitudes of the individual modes and the $e_{iojo}(s)$ are their unit polarization tensors. Supressing the two oo-components, these tensors can be written as 3x3 matrices labelled x y z in each index. They are, for a wave travelling in the +z-direction,

$$e(a) = \frac{1}{\sqrt{2}}\begin{pmatrix}1 & 0 & 0\\ 0 & -1 & 0\\ 0 & 0 & 0\end{pmatrix}, \quad e(b) = \frac{1}{\sqrt{2}}\begin{pmatrix}0 & 1 & 0\\ 1 & 0 & 0\\ 0 & 0 & 0\end{pmatrix}, \quad e(c) = \frac{1}{\sqrt{2}}\begin{pmatrix}1 & 0 & 0\\ 0 & 1 & 0\\ 0 & 0 & 0\end{pmatrix},$$

$$e(d) = \frac{1}{\sqrt{2}}\begin{pmatrix}0 & 0 & 0\\ 0 & 0 & 0\\ 0 & 0 & 1\end{pmatrix}, \quad e(e) = \frac{1}{\sqrt{2}}\begin{pmatrix}0 & 0 & 1\\ 0 & 0 & 0\\ 1 & 0 & 0\end{pmatrix}, \quad e(f) = \frac{1}{\sqrt{2}}\begin{pmatrix}0 & 0 & 0\\ 0 & 0 & 1\\ 0 & 1 & 0\end{pmatrix}. \tag{7.50}$$

The labels here correspond to the labels in Fig. 11 where it is shown what kind of displacement each mode induces on a sphere of test particles. The $e(a)$, $e(b)$ and $e(c)$ represent purely transverse modes,

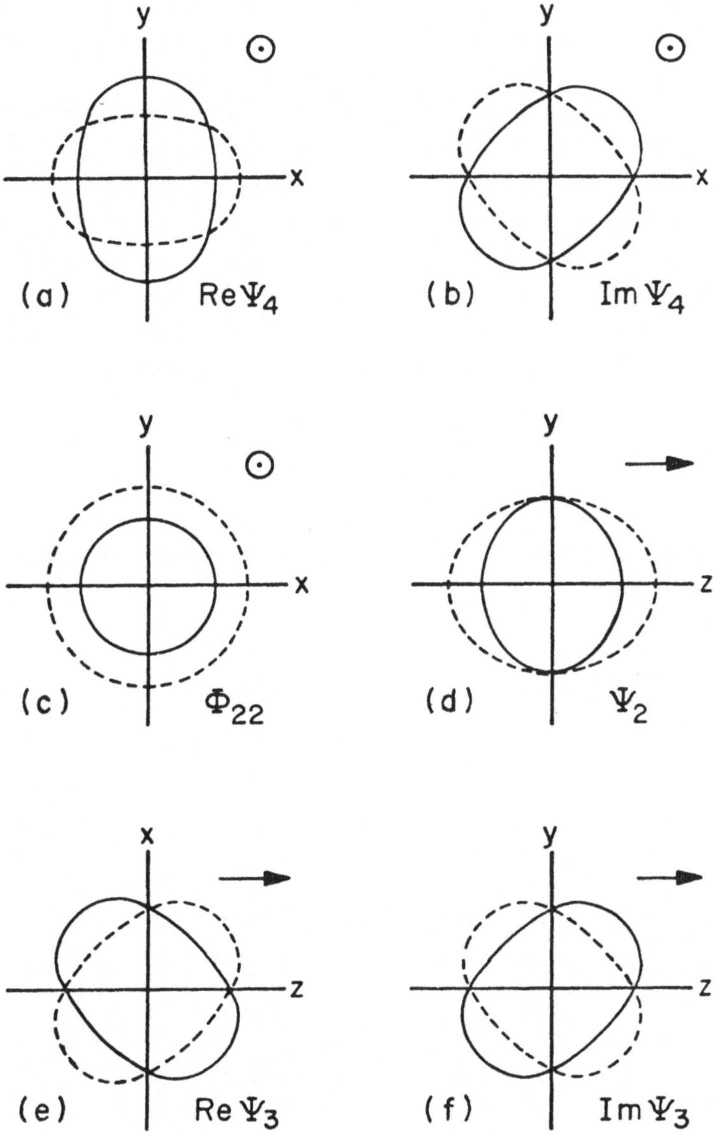

Figure 11. The six polarization modes of a weak, plane, null gravitational wave permitted in a generic metric theory of gravity. Shown is the displacement that each mode induces on a sphere of test particle ticles. The wave is propagating in the +z-direction (arrow at upper right) and has time dependence $\cos \omega t$. The solid line is a snapshot at $\omega t = 0$, the broken line at $\omega t = \pi$. There is no displacement perpendicular to the plane of the figure. The quantities given with the graph denote the Newman-Penrose quantities corresponding to each mode. (from Eardley et al. 1973).

the others, longitudinal or mixed modes. The $e(a)$ and $e(b)$ are present in General Relativity, and the scalar-transverse mode $e(c)$ occurs in addition in the Jordan-Brans-Dicke theory. The mode $e(d)$ is a scalar-longitudinal mode, and $e(e)$ and $e(f)$ induce oscillations in two orthogonal transverse-longitudinal planes (x-z and y-z). The longitudinal and mixed modes are characteristic of more general but still viable theories as listed in Eardley et al. (1973), but provide no completely reducible representation of the "little group" of Lorentz transformations (which leave one null vector invariant); these modes are therefore nonunitary and it is likely that theories allowing them cannot be quantized (Wagoner, IAU Symposium Nr. 64, 1973).

As there are six independent states of polarization, they may be represented by a 6x6 Hermitian polarization matrix Ξ_{AB} (A,B = 1,...,6). The trace of this matrix gives the energy flux of the wave. The individual components of Ξ_{AB} have to be obtained in the following manner. First define

$$J_{abcd\ efgh} = \frac{1}{16\pi\omega^2} (R_{abcd}\,\bar{R}_{efgh})_{av}\ , \tag{7.51}$$

derive from it the polarization tensor

$$J_{abcd} = J_{aebf\ cgdh}\, u^e u^f u^g u^h \tag{7.52}$$

for an observer with 4-velocity u^a (tetrad chosen as in (7.30), and project J_{abcd} onto three screens, namely the $e_{\hat{1}}-e_{\hat{2}}$, the $e_{\hat{1}}-e_{\hat{3}}$ and the $e_{\hat{2}}-e_{\hat{3}}$ planes:

$$\xi_{ijkl} \equiv J_{\hat{i}\hat{j}\hat{k}\hat{l}} = J_{abcd}\, e_{\hat{i}}{}^a e_{\hat{j}}{}^b e_{\hat{k}}{}^c e_{\hat{l}}{}^d . \tag{7.53}$$

The tensor ξ_{ijkl} satisfies

$$\xi_{ijkl} = \xi_{(ij)(kl)} = \xi_{klij}\ . \tag{7.54}$$

Due to (7.54) the following correspondence can be established between the six polarization states and the relevant (independent) index pairs $\{ij\}$ or $\{kl\}$ of ξ_{ijkl} :

$$\left\{ \begin{array}{c|cccccc} A & 1 & 2 & 3 & 4 & 5 & 6 \\ ij & 11 & 12 & 22 & 33 & 13 & 23 \end{array} \right\} \tag{7.55}$$

They can be collected in a column Ξ_A. Finally, the components of the polarization matrix are given by $\Xi_{AB} = (\Xi_A \bar{\Xi}_B)_{av}$.

The wave has six degrees of freedom labelled by the index pairs $\{ij\}$ in its potentials h_{ij}; hence $\Xi_A = h_{A,oo}$. The (real) diagonal elements of the coherence matrix are therefore

$$\begin{aligned} \Xi_{11} &= \xi_{1111}, \quad \Xi_{22} = \xi_{1212}, \quad \Xi_{33} = \xi_{2222}, \\ \Xi_{44} &= \xi_{3333}, \quad \Xi_{55} = \xi_{1313}, \quad \Xi_{66} = \xi_{2323}. \end{aligned} \tag{7.56}$$

The (complex) off-diagonal matrix elements are, naturally,

$$\Xi_{12} = \xi_{1112}, \quad \Xi_{13} = \xi_{1122}, \quad \Xi_{14} = \xi_{1133}, \text{ etc.} \tag{7.57}$$

All others are related by $\Xi_{AB} = \bar{\Xi}_{BA}$. As there are six real diagonal elements and 15 complex off-diagonal elements in the coherence matrix, this matrix is equivalently described by 36 Stokes parameters. The parameter S_o will again be the total energy flux of the wave,

$$\begin{aligned} S_o &= \mathrm{Tr}(\Xi\, \sigma_o) = \Xi_A{}^A \\ &= \xi_{1111} + \xi_{1212} + \xi_{2222} + \xi_{3333} + \xi_{1313} + \xi_{2323}, \end{aligned} \tag{7.58}$$

where σ_o is the 6x6 unit matrix. All other SP may be represented in the form $S_A = \mathrm{Tr}(\Xi\, \sigma_A)$ $(A=0,\ldots,35)$ provided one constructs the 35 remaining 6x6 Pauli matrices. They are the independent polarization states in a six-dimensional Hilbert space and the SP are their quantum-mechanical expectation values. The coherence matrix elements are expressed in terms of these as

$$\Xi_{12} = S_6 + i\, S_7, \quad \Xi_{13} = S_8 + i\, S_9, \quad \text{etc.} \tag{7.59}$$

The corresponding <u>degree of polarization</u> is defined by

$$d = \left[\sum_{i=0}^{35} s_i^2 \right]^{1/2} \qquad (7.60)$$

where $s_i = S_i/S_o$. If a wave is partially polarized, Ξ_{AB} splits into an unpolarized and a completely polarized part:

$$\Xi_{AB} = \frac{1}{2}(1-d)\,\delta_{AB} + d\cdot\Xi_A\cdot\Xi_B. \qquad (7.61)$$

7.7. Discussion

The idea of this chapter was to provide a formalism for the description of polarized weak electromagnetic and gravitational radiation in General Relativity. Such a formalism can be of use in the context of gravitational wave experiments, providing a framework analogous to the one already at the disposal of radio astronomers.

The question of polarization of gravitational waves will be an interesting feature when one is finally able to observe supernovae in the Virgo Cluster. As discussed by Eardley et al. (1973), the polarization mode occurring in a wave will also be crucial in distinguishing between the various viable theories of gravity.

This chapter was also concerned with the equation of radiative transfer for polarized radiation (in the high frequency limit). This can be of importance for the problem of electromagnetic radiation emitted near neutron stars (pulsars) in collapsed objects (black holes). The transfer equation for the gravitational Stokes parameters may be useful in discussing the problem of interaction of gravitational waves with other types of matter such as elastic bodies (neutron stars); for this question the formalism developed by Carter & Quintana (1972) and Carter (1973) may be used to obtain the equation of radiative transfer in the presence of an elastic body.

VIII. GEODESIC SYNCHROTRON RADIATION

8.1 Introduction

In this chapter properties of radiation emitted by particles moving on high-energy geodesic orbits will finally be discussed. In Chap. III it was found that particles which orbit in the vicinity of the closed circular photon orbits (of the Schwarzschild or Kerr geometry) have large local energy; but it was also noted there that these orbits are unstable and hence irrelevant to astrophysical applications. For particles moving in stable circular orbits, the main radiation is in the low-frequency (fundamental) mode, for which numerical computations are required (Davis et al. 1972). In Sec. 8.3 an argument is given proving this fact.

The high-frequency radiation, to which the methods of this chapter apply, is a major part of the radiated energy only in the case when the particle moves on relativistic circular geodesics. Few other examples have been given in the theory of general relativity for which the radiation by a well-defined source is computed. One such calculation is Einstein's well-known computation of the gravitational radiation emitted by a nonrelativistic oscillating quadrupole in the linearized theory of gravity. Other calculations include the radiation from a vibrating neutron star (Thorne 1970), gravitational bremsstrahlung (Peters 1970, 1972) and the radiation emitted by a small mass falling radially into a Schwarzschild black hole (Davis et al. 1970) ("plunge radiation"). However, neither of these examples cited - and this includes GSR - provide exact models for gravitational radiation linking radiation with its source, since both source and radiation field are treated in the linearized approximation.

Another point concerns the question whether a particle in geodesic motion can radiate at all and the relation of this problem to the equivalence principle. (For flat spacetime treatments of this question see Bondi & Gold 1955; Fulton & Rohrlich 1960; Rohrlich 1963). A test particle which moves on a geodesic does not radiate at all because of the very axiomatic construction of a test particle. If test particles are used as linearized sources for the computation of linearized radiation fields the geodesic motion can only be considered as an approximation during the radiation process and the equivalence should not be applied outside its domain of validity. This approximation holds as long the radiation damping perturbs the world line only slightly (Zel'dovich & Novikov 1971). In fact, the assumption of geodesic motion for the computation of GSR provides another conceptual limitation for the computations done in this chapter. This is in addition to the limitations implied by the test particle assumption and the use of linearized field equations.

These assumptions together with the high-frequency approximation allow the differential equations to be solved analytically in a WKB approximation. It will be shown that of the high frequency modes those with $\ell \neq m+s$ contribute negligibly to the radiation energy. Thus, the relations

$$M\omega \gg 1 \quad (\Rightarrow m \gg 1) \quad \text{and} \quad \left| \left[\ell \pm (m+s)\right]/m \right| \ll 1 \tag{8.1}$$

are the limits defining the synchrotron approximation of this chapter.

To warm up mathematically first the simplest case is considered, namely scalar radiation emitted by a particle moving in a circular geodesic orbit about a Schwarzschild black hole. Subsequently, the high-frequency approximations of radial & angular functions are derived for the Kerr geometry together with the frequency spectra for scalar, electromagnetic and gravitational radiation.

8.2 Scalar GSR in the Schwarzschild Geometry

Consider the emission of scalar waves by a test particle in a relativistic circular orbit in the Schwarzschild geometry as described in Chap. III (Breuer, Chrzanowski, Hughes III and Misner 1973). The results of Chap. VII specialized to the case $a=0$ will be used. The interaction between a scalar field ϕ and a test particle of mass μ is described by the action

$$I = -\frac{1}{8\pi}\int d^4x\sqrt{-g}\;\phi_{,a}\phi^{,a} - \mu\int d\tau\,(1+f\phi)\,(-\dot{z}^a\dot{z}_a)^{1/2}. \tag{8.2}$$

The test particle follows the world line $z^a(\tau) = (t,R,\Theta,\Phi)$ with affine parameter τ, which may be chosen to be the proper time. The four velocity is $\dot{z}^a = u^a = u^o(1,\dot{R},\dot{\Theta},\dot{\Phi})$, $u^o \equiv dt/d\tau$. The constant f measures the strength of the "scalar charge" and can again be chosen to be $f = \sqrt{G}$. If there is no background scalar field then no other terms enter into the source terms to first order and the scalar field density can be neglected as it is of second order. By standard methods the following energy-momentum tensor is obtained

$$\begin{aligned} T^{(o)}_{ab} = {} & \frac{1}{4\pi}\left(\phi_{,a}\phi_{,b} - \frac{1}{2}g_{ab}\,\phi_{,c}\phi^{,c}\right) \\ & + (-g)^{-1/2}\int d\tau(1+f\phi)\,\delta^4(x-z)\,\dot{z}_a\dot{z}_b\,. \end{aligned} \tag{8.3}$$

By varying (8.2) with respect to ϕ the field equation for ϕ is found to be

$$\frac{\partial}{\partial x^a}\left(\sqrt{-g}\,g^{ab}\frac{\partial\phi}{\partial x^b}\right) = 4\pi f\sqrt{-g}\,T. \tag{8.4}$$

For the homogeneous case this equation was first separated by Wheeler (1962) and studied by Matzner (1968), Vishveshwara (1970), and Price (1972) for the problem of scattering of massless scalar waves by a Schwarzschild black hole. For a particle moving in a circular orbit with radius r_o and angular velocity ω_o (see (3.13)), one may write

$$\delta^3\left[x-z(\tau)\right] = \delta(r-r_o)\,\delta(\theta-\pi/2)\,\delta(\varphi-\omega_o t). \tag{8.5}$$

Then

$$\sqrt{-g}\,T = \mu\int d\tau\, u_a u^a\,\delta^4[x-z(\tau)] = -\mu\left(\frac{dt}{d\tau}\right)^{-1}\delta^3[x-z(\tau)]\,.$$

Exploiting the spherical symmetry of the Schwarzschild metric the scalar field may be written in separated form as follows:

$$\phi = \sum\!\!\!\!\!\!\int\; r^{-1}\,{}_o u_{\ell m\omega}(r)\;Y^m_\ell(\theta,\varphi)\,e^{-i\omega t}\,, \tag{8.6}$$

where $\displaystyle \oint = \int_{-\infty}^{\infty} d\omega \sum_{\ell=0}^{\infty} \sum_{m=-\ell}^{+\ell}$. The reality conditions are given by

$$ {}_0u_{\ell\,-m\omega} = (-1)^m \bar{u}_{\ell\,m\omega}; \qquad Y_\ell^{-m} = (-1)^m \bar{Y}_\ell^m \quad . \tag{8.7}$$

Insertion of ϕ into the scalar equation (8.4) leads to a radial equation

$$\left[- \frac{d^2}{dr^{*2}} + {}_0V - E \right] {}_0u_{\ell m\omega}(r) = 4\pi f \int \sqrt{-g}\, d\Omega\, T\, \bar{Y}_{\ell m} e^{i\omega t} , \tag{8.8}$$

where

$$ {}_0V(r) = (1 - \frac{2M}{r}) \left[\frac{2M}{r^3} + \frac{\ell(\ell+1)}{r^2} \right],$$

$$ E = \omega^2 = m^2 \omega_0{}^2 = M\, m^2/r_0{}^3 \ ,$$

and $r^* = r - 3M + 2M\, \ln(r/M-2)$. In the limit $\ell \gg 1$ the potential becomes

$$ {}_0V(r) = (1 - \frac{2M}{r}) \frac{\ell(\ell+1)}{r^2} + O(\ell^0) . \tag{8.9}$$

The effective potential is plotted in Fig. 12 in its asymptotic form for large ℓ. The solution of Eq. (8.8) is given by

$$ {}_0u_{\ell m\omega}(r) = 2f \int \sqrt{-g}\, d^4x_0 G(r,r_0) \bar{Y}_{\ell m}(\theta_0, \varphi_0) e^{i\omega t_0} . \tag{8.10}$$

The Green's function $G(r,r_0)$, which is a solution of the equation

$$\left[- \frac{d^2}{dr^{*2}} + {}_0V - E \right] G(r^*, r^*_0) = \delta(r^* - r^*_0) \ , \tag{8.11}$$

is obtained by matching at r^*_0 two solutions of the homogeneous equation satisfying different boundary conditions, as discussed in Chap.VII. Setting $a=0$ in Eq. (6.38) one has

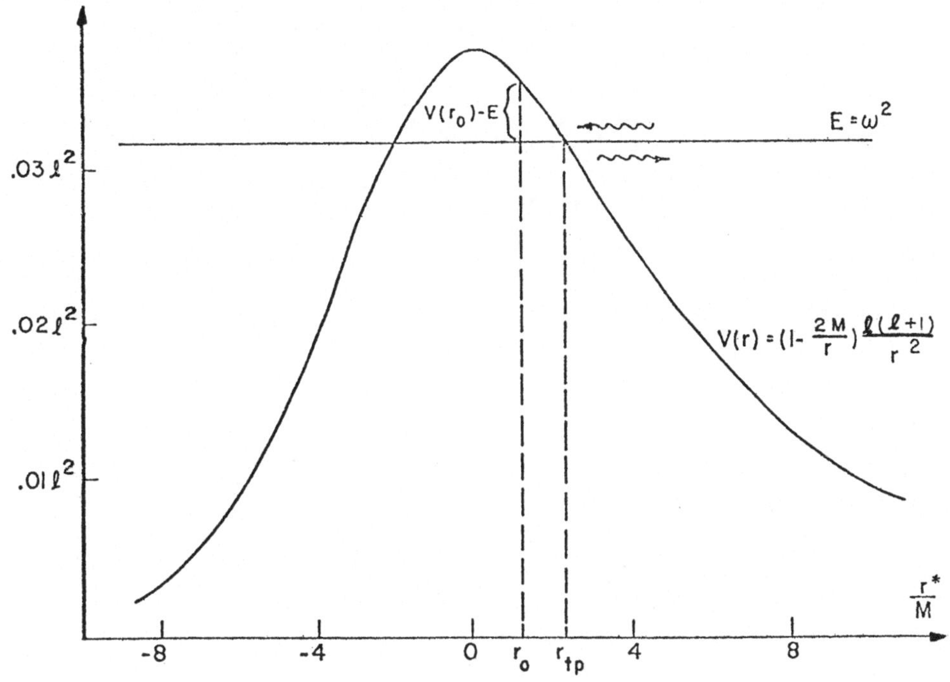

Figure 12. Reduced ($\ell \to \infty$) effective potential for radiation in the vicinity of a Schwarzschild black hole. This potential tends to zero for $r^* \to \pm\infty$. The radial coordinate is scaled such that a maximum is reached at $r^* = 0$ or $r = 3M$, where photons with impact parameter $b = 3\sqrt{3}\,M$ remain in an unstable circular orbit. Generally, $r_o^* < r_{tp}^*$ and hence the potential barrier factor $\int_{r_o^*}^{r_{tp}^*} [V(r_o) - E]\, dr^*$ is large. Only in the synchrotron limit (8.1) one finds $r_{tp} \to r_o \simeq 3\,M$ whence the barrier becomes small and high multipole moments can be excited.

$$ {}_{o}u^{up} \sim |\omega|^{-1/2} \begin{cases} e^{i\omega r^*} & , \quad r^* \to +\infty, \\ \mathcal{T}^{-1}\, e^{i\omega r^*} - \overline{(S/\mathcal{T})}\, e^{-i\omega r^*}, & r^* \to -\infty, \end{cases} \tag{8.12a} $$

$$ {}_{o}u^{in} \sim |\omega|^{-1/2} \begin{cases} e^{-i\omega r^*} + S\, e^{i\omega r^*} & , \quad r^* \to +\infty, \\ \mathcal{T}\, e^{-i\omega r^*} & , \quad r^* \to -\infty, \end{cases} \tag{8.12b} $$

For $r^* > r^*_o$ the resulting Green's function becomes

$$ G(r^*, r^*_o) = \frac{i}{2} \frac{\omega}{|\omega|}\, {}_{o}u^{in}(r^*_o)\, {}_{o}u^{up}(r^*) $$

and the corresponding field becomes

$$ \phi(x) = \mathop{\not\Sigma}\!\!\!\!\int i \frac{\omega}{|\omega|}\, \phi^{up}(x, \ell m\omega) \left\langle \phi^{out}(x, \ell m\omega),\, fT \right\rangle , \tag{8.13} $$

where

$$ \phi^{up}(x, \ell m\omega) = \frac{{}_{o}u^{up}_{\ell m\omega}(r)}{r}\, Y_{\ell m}(\theta,\varphi)\, e^{-i\omega t}, $$

$$ \phi^{out}(x, \ell m\omega) = \frac{{}_{o}u^{out}_{\ell m\omega}(r)}{r}\, Y_{\ell m}(\theta,\varphi)\, e^{-i\omega t} . \tag{8.14} $$

The total power radiation "out" to $\mathcal{J}^+$ or "down" across the future event horizon is obtained by integrating over solid angles the outward (or inward) flux of energy. The result is

$$ E^{(o)}_{out} = -\lim_{t\to+\infty} \int d\Omega\, r^2\, T^r{}_t = -\frac{1}{4\pi} \int d\Omega\, r^2\, \phi^{,r}\, \phi_{,t} $$

$$ = \lim_{t\to+\infty} \int d\Omega\, dt\, \frac{r^2\omega^2}{4\pi}\, |\phi|^2 . \tag{8.15} $$

Applying formulae (8.15) to the solution (8.13) the energy spectrum $dE^{(o)}_{out}/d\omega$ can be defined by

$$ E^{(o)}_{out} \equiv \int d\omega \frac{dE^{(o)}_{out}}{d\omega} = \mathop{\not\Sigma}\!\!\!\!\int \omega \left| \left\langle \phi^{out}(x, \ell m\omega), fT \right\rangle \right|^2 . \tag{8.16} $$

The scalar power formula follows from (8.15) by assuming that the particle radiates in the interval $-T \leq t \leq T$, where $T \gg M$; then

$$E^{(o)}_{out} = \int_{-T}^{T} dt\, P^{(o)}_{out} = 2\pi P^{(o)}_{out} \quad \text{and} \quad P^{(o)}_{out} = \int_0^\infty d\omega \frac{dP^{(o)}_{out}}{d\omega} .$$

Therefore, for the specific source term (8.5) one has

$$\frac{dP^{(o)}_{out}}{d\omega} = \sum_{\ell=m}^{\infty} \frac{2 f^2 \mu^2 m}{(dt/d\tau)^2 r_o^2} \left| {}_o u^{out}_{\ell m \omega}(r_o) \right|^2 \left| Y_{\ell m}(\pi/2, 0) \right|^2$$

$$P^{(o)}_{out} = \sum_{m=0}^{\infty} \frac{dP^{(o)}_{out}}{d\omega} \Delta\omega \simeq \int d\omega \frac{dP^{(o)}_{out}(\omega)}{d\omega} \omega_o , \tag{8.17}$$

with $\omega_o = \Delta\omega$ being the frequency interval between modes m and $m+1$. A WKB solution for ${}_o u^{out}$ may be found if the potential satisfies the necessary condition

$$\frac{1}{2} \left| \frac{\frac{d}{dr^*}(V-E)}{(V-E)^{3/2}} \right|_{r=r_o} \ll 1. \tag{8.18}$$

In Chap. III it was shown that the particle is relativistic only if $r_o = r_\gamma(1+\delta) = 3M(1+1/9\gamma^2)$ and $\gamma^2 \gg 1$. Hence for large m the leading terms for numerator & denominator of (8.18) become respectively

$$\frac{d}{dr^*}(V-E) \Big|_{r=r_o} = - \frac{2Mm^2}{k r_o^4 \gamma^2}, \quad V(r_o) - E = \frac{m}{27M^2}\left[1 + 2k + \frac{m}{3\gamma^2}\right], \quad k = \ell - m .$$

Substituting in (8.18), yields the following condition for a WKB approximation to be valid:

$$m^{1/2} \left[\sqrt{3}\, \gamma^2 (1+2k+m/3\gamma^2)\right]^{-1} \ll 1. \tag{8.19}$$

This obviously holds for all m if $\gamma^2 \gg 1$. If one defines the barrier penetration factor $\Theta(r_o)$ and the quantity $K(r)$ by

$$\Theta(r_o) = \int_{r_o^*}^{r^*_{tp}} \left[V(r) - E\right]^{1/2} dr^* \equiv \int_{r_o^*}^{r^*_{tp}} K(r)\, dr , \tag{8.20}$$

then the WKB solution for ${}_o u^{up}$ is given by

$$ {}_o u^{up}_{\ell m\omega}(r^*_o) = e^{-i\pi/4}\, K^{-1/2}\,(r^*_o)\, \exp\left[-\Theta(r_o)\right] . \tag{8.21} $$

The outer classical turning point r_{tp} is defined through $V(r_{tp}) = E$. The barrier penetration factor has to be determined for large ℓ, i.e. $r_o \simeq r_\gamma$ near the peak of the potential. There the potential is approximated by a parabola, namely by

$$ \begin{aligned} V(r) &= V(r_\gamma) + \frac{1}{2}\frac{d^2}{dr^{*2}} V(r_\gamma)\cdot(r^*-r^*_\gamma)^2 \\ &= \frac{m}{27M^2}\left[1 + 2k + \frac{m}{3\gamma^2}\right] - \left(\frac{m}{27M^2}\right)^2 (r^* - r^*_\gamma)^2 . \end{aligned} \tag{8.22} $$

Integration yields

$$ \Theta(r^*_o) = \frac{\pi}{4}\left[1 + 2k + m/3\gamma^2\right] \equiv \frac{\pi}{4}\varepsilon . \tag{8.23} $$

This barrier factor determines which modes ℓ, m contribute significantly to the radiation. One may introduce certain critical cut-off values k_{crit} and $m_{crit} = \omega_{crit}/\omega_o$ beyond which this contribution is, say, less by a factor e^2 than the leading term. They are defined by

$$ \begin{aligned} \Theta(k=0\ ,\ m=m_{crit}) &= \Theta(k=0,\ m=0) + 1\ , \\ \Theta(k=k_{crit},\ m) &= \Theta(k=0,\ m) + 1\ . \end{aligned} \tag{8.24} $$

With the aid of (8.23) one obtains the values

$$ \begin{aligned} & m_{crit} = \frac{12}{\pi}\gamma^2 = \frac{4}{\pi\delta}\ , \qquad k_{crit} = 0, \\ \Rightarrow \qquad & \varepsilon = 1 + 2k + \frac{4}{\pi}\frac{m}{m_{crit}}\ . \end{aligned} \tag{8.25} $$

Relations (8.25) imply that nearly all of the power is radiated into the $\ell = m$ mode. Finally, Stirling's formula is used to obtain the asymptotic form of the $Y_{\ell m}(\theta,\varphi)$:

$$Y_{\ell=m+k,m}(\theta,\varphi) = \begin{cases} (-1)^m (4\pi^3)^{-1/4} \dfrac{(k!)^{1/2} m^{1/4}}{(\frac{k}{2}!)\ 2^{k/2}} \sin^m\theta\ e^{im\varphi}(1+\frac{1}{4m}),\ k \text{ even}, \\ (-1)^m\ \pi^{-3/4} \dfrac{[(k-1)!]^{1/2}\ m^{3/4}}{(\frac{k-1}{2})!\sqrt{2}^{k-1}} \sin^m\theta \cos\theta\ e^{im\varphi}(1+\frac{11}{8m}),\ k \text{ odd}. \end{cases} \quad (8.26)$$

Inserting ${}_ou^{up}$ and $Y_{\ell m}$ into (8.17) and setting k=0 leads to the following formula for the power spectrum

$$\frac{dP_{out}^{(o)}}{d\omega} = \left(\frac{f\mu}{M}\right)^2 \frac{1}{27\pi^{3/2}\omega_o} \varepsilon^{-1/2} \left(\frac{\omega}{\omega_{crit}}\right) \exp(-\pi\varepsilon/4) \quad (8.27)$$

where $\varepsilon = 1 + \frac{4\ m}{\pi\ m_{crit}}$. The power spectrum is proportional to the frequency below $\omega_{crit} = \omega_o m_{crit}$ and exponentially damped above ω_{crit} (see Fig. 13).

To calculate the total radiated, Eq. (8.27) must be summed over m. As noted in the caption to Fig. 13, this summation can be converted to an integral over frequency. Numerically one finds that

$$P_{out}^{(o)} = \left(\frac{f\mu}{M}\right)^2 3.9\text{x}10^{-3} \times \gamma^2 . \quad (8.28)$$

Here $P_{out}^{(o)}$ is dimensionless and should be multiplied by $c^5/G = 3.65 \times 10^{59}$ erg/sec $= 2.03 \times 10^5\ M\ c^2$/sec to obtain the corresponding value in CGS units.

At fixed frequency $\omega = m\omega_o$, the angular distribution of the radiation is easily computed, since only a single conjugate pair of terms in the series expansion (8.27) is significant, namely, that with $\pm m = \ell = |\omega/\omega_o|$. The amplitude ϕ is therefore proportional to $Y_{mm}(\theta,\varphi)$ and for the radiated power spectrum per unit solid angle one consequently has

$$\frac{d^2}{d\Omega\, d\omega} P_{out}^{(o)} = \frac{dP_{out}^{(o)}}{d\omega} \left| Y_{mm}(\theta,\varphi) \right|^2 . \quad (8.29)$$

Or, using Eq. (8.26), one gets

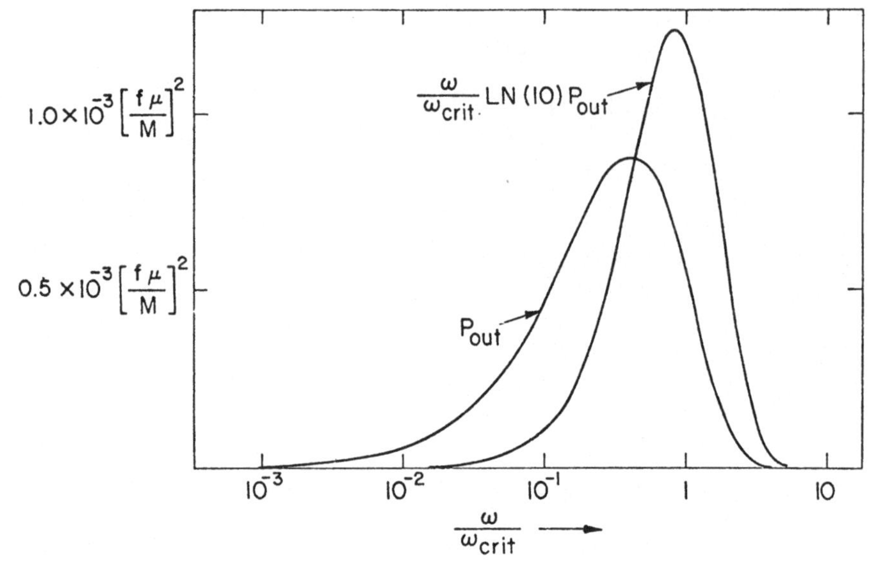

Figure 13. Scalar power for GSR in the Schwarzschild geometry in the limit $\delta = (3\gamma^2)^{-1} \ll 1$. Here

$$P_{out}^{(o)}(m) = \sum_{\ell=m}^{\infty} P_{out}^{(o)}(\ell,m)$$

for scalar radiation is given as a function of $\omega/\omega_{crit} = \frac{\pi}{4} m\delta$ for each frequency harmonic $m = \omega/\omega_o \gg 1$. The total power emitted in all high harmonics is

$$P_{out}^{(o)} = \sum_{m=0}^{\infty} P_{out}^{(o)}(m) \simeq \int dm\, P_{out}^{(o)}(m) .$$

Since this can be written as $P_{out}^{(o)} = \int (2\pi/\omega_o) P_{out}^{(o)}(\omega)\, d\omega$, the power emitted per unit frequency is $dP_{out}^{(o)}/d\omega = \frac{2\pi}{\omega_o} P_{out}^{(o)}$. To obtain $P_{out}^{(o)}$ in terms of the area under a curve in a semilog plot, where the abscissa is really $\log_{10}(\omega/\omega_{crit})$, one writes

$$P_{out}^{(o)} = \frac{\omega_{crit}}{\omega} \int_0^{\infty} (\ln 10)\, P_{out}^{(o)}(\omega)\, d(\log_{10} \frac{\omega}{\omega_{crit}}) .$$

The integrand is plotted to show that the bulk (82 %) of the energy is emitted in a decade

$$10^{-0.7} < \omega/\omega_{crit} < 10^{+0.3} .$$

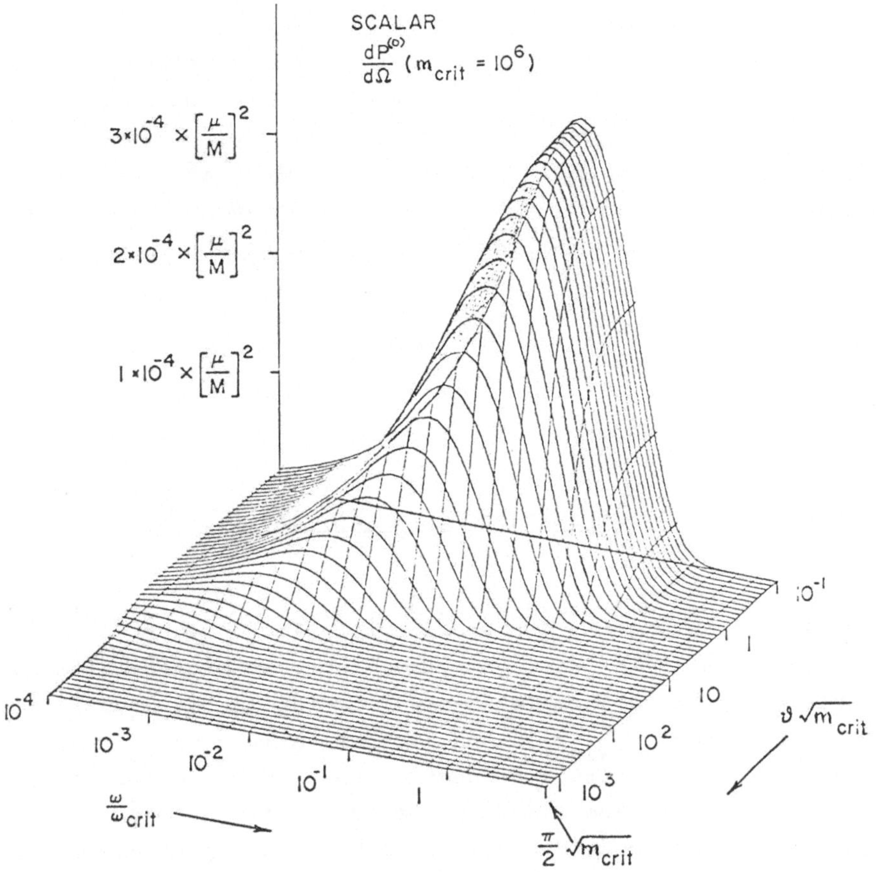

Figure 14. Scalar power per unit solid angle $\vartheta \equiv \pi/2 - \theta$ (θ is the polar angle) and frequency. The scaling of the logarithmic ϑ-axis is chosen so that the half width of the beam is at $\vartheta\sqrt{m}_{crit} = 1$ for $\omega/\omega_{crit} = 1$.

$$\frac{d^2}{d\Omega\, d\omega} P^{(o)}_{out} = \left(\frac{f\mu}{M}\right)^2 \frac{1}{16\pi^3} \left(\frac{m}{m_{crit}}\right) \varepsilon^{-1/2}\, e^{-\pi\varepsilon/4}\, m^{1/2} \cos^{2m}\vartheta , \quad (8.30)$$

where $\vartheta = \pi/2 - \theta$. A plot of (8.30) as a function of both ϑ and ω is shown in Fig. 14. The function $\cos^{2m}\vartheta$ is harply peaked about $\vartheta = 0$ for $m \gg 1$ so that the radiation is strongly peaked with respect to the plane to the orbit. From Eq. (8.30) the half-width $\Delta\vartheta$ of the beam is easily calculated; for small ϑ one has $\cos^{2m}\vartheta \approx \exp(-\frac{1}{2}\vartheta^2 |m|)$. The angle for which the power e-folds is therefore given by

$$\Delta\vartheta = |m|^{-1/2} = |\omega/\omega_o|^{-1/2} . \quad (8.31)$$

8.3 High-Frequency Approximation of ${}_su(r)$ and ${}_sS(\frac{\pi}{2},0)$ in the Kerr Geometry

Before attacking the GSR problem for electromagnetic and gravitational radiation in the Kerr geometry, two more prerequisites are required. First, starting from Teukolsky's radial master equation (6.30), the WKB solution for ${}_su_{\ell m\omega}(r_o)$ and its derivatives have to be obtained. It will also be shown in this connection that only negligible synchrotron radiation can be expected from stable orbits. Secondly, since only equatorial orbits are considered, the high-frequency expansion of spin-weighted spheroidal harmonics and their derivatives have to be calculated only in the equatorial plane. Although the previously obtained high-frequency expansion could be used for this pupose, it turns out that, to the accuracy required, straightforward application of methods used by Chrzanowski & Misner (1974) yields the desired results in a simpler fashion.

One looks for a WKB solution of the equation

$$\left[-\frac{d^2}{dr^{*2}} + {}_sV(r)\right] {}_su(r^*) = (r^2+a^2)^{-3/2}\, \Delta_K^{1+2/2}\, {}_sT , \quad (8.32)$$

where

$$ {}_sV(r) = -\omega^2 + (r^2+a^2)^{-1}\left[2am\omega + 2is(r-M)\omega\right] - \tag{8.33}$$

$$ - (r^2+a^2)^{-2}\left[a^2m^2 + 2is(r-M)am + \Delta_K(4ir\omega s - {}_s\lambda - s) - s^2(r-M)^2\right] + $$

$$ + (r^2+a^2)^{-3}\,\Delta_K(3r^2+a^2-4Mr) - (r^2+a^2)^{-4}\,3r^2\Delta_K^2 \quad , $$

where

$$ {}_s\lambda = {}_sA + a^2\omega^2 - 2am\omega \equiv {}_s\Omega + m^2 + a^2\omega^2 - 2am\omega \; . $$

In the high-frequency limit, $\omega^2 \to \pm\infty$, the potential reduces to

$$ {}_sV(r) = -\omega^2 + \frac{4Mar\omega m - a^2m^2 + ({}_s\Omega + m^2 + a^2\omega^2)\Delta_K}{(r^2+a^2)^2} - $$

$$ - \frac{2is}{r^2+a^2}\left[(r+M)\omega + \frac{(r-M)am - 4Mr^2\omega}{r^2+a^2}\right] + O(\omega^0) \; . \tag{8.34}$$

There are two remarkable features of the potential (8.33). First, it is complex and the leading term in its imaginary part is of second order in frequency. Secondly, this term is precisely the leading term involving spin. In addition, in the real part of the potential the spin enters indirectly through the eigenvalues ${}_s\lambda$ appearing in the term of third order in ω .

In general, not much is known about wave equations with complex potentials (cf. Alfaro & Regge 1965). For axisymmetric perturbations (m=0) of the Kerr black hole, Chandrasekhar & Detweiler (1975) have studied the complex nature of this potential. There they derive various formulations for the (specialized) potential but show that irrespective of whether or not the potentials are real or complex (through ω) they all lead to identical reflection and transmission coefficients. For the purpose of GSR only real frequencies will be considered.

However, there is no intrinsic physical significance contained in the imaginary part of the potential ${}_sV(r)$. Indeed, this is already obvious from considering the Schwarzschild limit, where this potential is still complex while the Regge-Wheeler and Zerilli-equations contain only real effective potentials. Evidently, the presence of the imaginary term is related to the different asymptotic behaviour of the NP-fields when approaching $\mathcal{J}^+$ and $\mathcal{J}^-$, respectively. In fact, Detweiler (1975) has shown that, by choosing suitable combinations of radial functions of positive and negative spin weight, a radial equation results with a completely real effective potential. This corresponds to decoupling and separating not for the NP outgoing or ingoing radiation fields themselves but for a certain linear combination of NP-fields from the start. This combination is more directly related to the energy flux and contains in- and outgoing radiation.

If one splits the potential (8.34) in an obvious manner into the real and imaginary part as ${}_sV = \omega^2(V_0 + is\, V_1)$, then the standard WKB approximation yields (up to a constant phase)

$$ {}_su^{in}(r_o^*) \simeq \left[\omega K(r_o^*)\right]^{-1/2} \exp\left[-{}_s\theta(r_o^*)\right] , \tag{8.35}$$

where

$$ {}_s\Theta(r_o^*) = \int_{r_o^*}^{r_{tp}^*} dr^*\left[\omega K(r) + i\,\frac{s}{2}\,\frac{V_1(r)}{K(r)}\right] ; \quad K(r) = \left[V_o(r)\right]^{1/2} . $$

Note, that the imaginary part $2is(r-M)/(r^2+a^2)$ provides the correct asymptotic radial behaviour for ${}_s\bar{u}^{in} \sim r^{\pm s} e^{\pm i\omega r}$.

No GSR from stable orbits

Now it will be demonstrated that only a negligible amount of synchrotron radiation can be expected from particles in stable orbits. A necessary condition for GSR in which a large amount of power is radiated at high frequencies ($\omega \gg M^{-1}$), is that ${}_s\Theta(r_o^*) \ll 1$ in Eq. (8.35). Clearly, this condition can only be satisfied if ${}_sV(r_o^*) \ll 1$. Inspecting the potential in (8.32) and using Eq. (3.26), it becomes obvious that high-frequency radiation may only occur for $r_o \simeq r_\gamma$. In BL coordinates the radius $r_{\ell s}$ of the last stable orbit approaches r_γ only when $a \rightarrow M$; consequently, $a^2 \simeq M^2$ is the only possible candidate for GSR among stable circular obits. When $a^2 = (1-\alpha^2)M^2$, $\alpha^2 \ll 1$, according to Chap. III $r_{\ell s}$ and the edge of the barrier r_{tp} are given by

$$r_{\ell s} = \left[1 + (2\alpha^2)^{1/3}\right] M, \quad r_{tp} = \left[1 + \tfrac{3}{2}(2\alpha^2)^{1/3}\right] M, \tag{8.36}$$

respectively. By approximating the potential by a parabola about r_γ, one gets

$$_s\Theta(r_o^*) = \frac{\sqrt{3}}{2}\, m\, \omega_o M \int_{1/2}^{3/2} dx\, x^{-2} \left[(x-1)(\tfrac{3}{2}-x)\right]^{1/2} + \mathcal{O}(s\omega^2)$$

$$\simeq 0.12\, m\, \omega_o M. \tag{8.37}$$

Thus, for $m \gg 1$, independently of spin, the amount of synchrotron radiation from stable circular orbits is negligible.

No GSR from plunge orbits

Another class of stable orbits which may be envisaged as astrophysically plausible sources for GSR are those of particles "plunging" more or less radially into a black hole. Test particles falling into a Schwarzschild black hole from rest at past timelike infinity I^- have been considered by Davis et al. (1972). They found the total amount of gravitational energy radiated to $\mathcal{J}^+$ is given by

$$E_{out}^{(2)} = 0.0104\, \mu^2 c^2 \left(\frac{\mu}{M}\right) . \tag{8.38}$$

Further, this energy is radiated primarily into the fundamental quadrupole mode, and higher modes are damped exponentially. Chrzanowski (1973) has studied relativistic plunge orbits in both Schwarzschild and Kerr geometries and, qualitatively, obtained the following results. For such relativistic "dive-in" particles with less angular momentum L_z than required for circular motion (including "radial" motion with $L_z = 0$) there is, in fact, substantial synchrotron radiation when the particles approach the horizon; however, the radiation is beamed in the forward direction with beaming angle γ_B^{-1} (see Chap. III). For synchrotron models null orbits wihich could "trail off" to $\mathcal{J}^+$ do not lie within this forward cone and only the less focussed, low mode radiation is

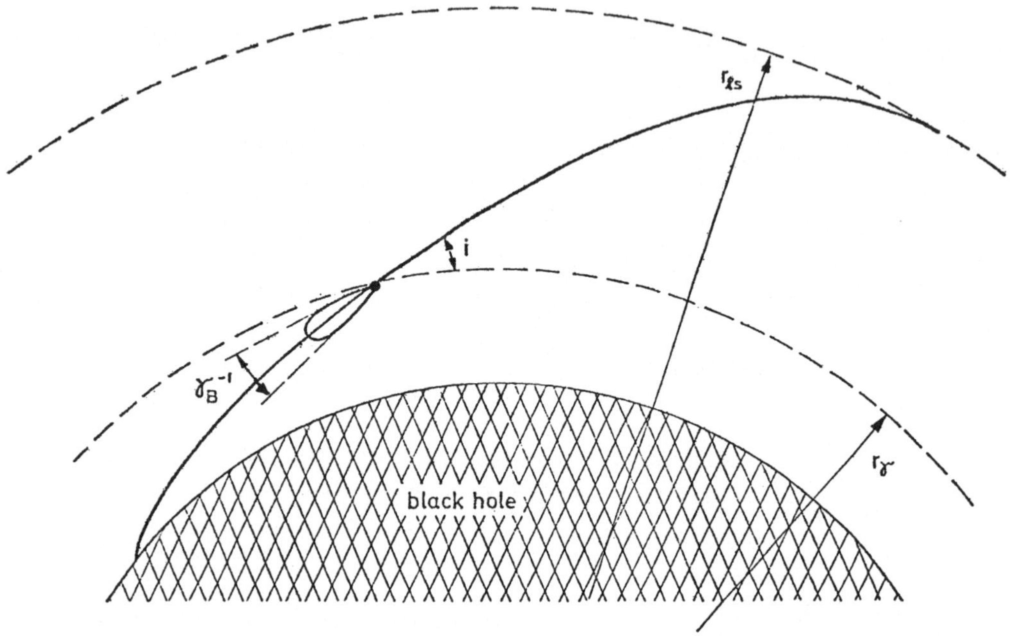

Figure 15. (from Chrzanowski 1973). Illustration of the situation that arises in the plunge model. Consider a particle on the last stable circular orbit at radius $r_{\ell s}$ about an extreme Kerr black hole. After a slight perturbation, the particle falls off the stable orbit and spirals along the solid line into the black hole. According to the locally non-rotating observers, the particle is highly relativistic ($p_{\hat{t}} \gg \mu$) as it nears the radius of the photon circular orbit r_γ if $M - a \ll M$. Hence, the radiation emitted in this region can be expected to be strongly beamed and describable as synchrotron radiation. The relevance of the plunge model for GSR hinges on the answer to the question: is the null circular orbit, describing radiation which can escape to infinity, within the forward beaming cone of the particle? That is, does the pitch angle i of the spiral depicted in the deawing satisfy $i < \left[\frac{p_{\hat{t}}}{\mu}\right]^{-1}$? Chrzanowski (1973) concludes that the answer is no.

able to reach a distant observer (outside the black hole, that is). The bulk of the radiation gets absorbed by the black hole (see Fig. 15).

High-frequency solution for ${}_s u(r_o)$

Since the barrier penetration factor must obey ${}_s\theta(r_o) \ll 1$ at high frequencies to allow GSR, it must be demonstrated that the condition (prime denotes d/dr^*)

$$\left|\frac{\omega K' + sV_1}{\omega^2 K^2}\right|_{r=r_o} \leq \left|\frac{V_o'}{2\omega^{1/2}\, V_o^{3/2}}\right| + s\left|\frac{V_1}{\omega^2\, V_o}\right|_{r=r_o} \ll 1 \qquad (8.39)$$

is satisfied for the WKB approximation (8.35) to be valid. GSR will only occur when $r_o \simeq r_\gamma$. Using Eq. (3.26) for r_γ, which in this limit also holds for r_o, the potential ${}_sV(r_o)$ can be shown to be

$$ {}_sV(r) \simeq (r^2+a^2)^{-2}\left[\Delta_K({}_sQ+\omega^2 r_o{}^3/M) - (r^2-a\, r_o{}^{3/2}/M^{1/2})^2\omega^2 + \right. $$
$$ \left. - 2is(r+M)(r^2+a^2)\omega\right]. \qquad (2.40)$$

Furthermore, one finds that for $r_o \simeq r_\gamma$

$$ {}_sV(r_o) \simeq \omega^2\left[\frac{\Gamma^4 W - 2is(r+M)(r^2+a^2)\frac{1}{\omega}}{(r^2+a^2)^2}\right]\Bigg|_{r=r_o}, $$
$$ V_o'(r_o) \simeq -\frac{\Delta_K(r^2-a^2)(r-M)^2 r^2}{2(r^2+a^2)^4\,\gamma^2 M}, \qquad (8.41)$$

where

$$ W = \frac{\sqrt{3}\,\Delta_\gamma}{\omega\, r_\gamma^3}\,\varepsilon, \quad \varepsilon = 1 + 2k + \frac{4}{\pi}\,\frac{m}{m_{crit}}, \quad \Delta_\gamma = r_\gamma{}^2 - 2Mr_\gamma + a^2, $$

$$ m_{crit} = \frac{2\sqrt{3}}{\pi}\,\frac{r_\gamma + 3M}{(r_\gamma M)^{1/2}}\,\gamma^2 \, . \qquad (8.42)$$

Substituting back into (8.39), the WKB approximation is valid if

$$m^{-1/2}\left[\left\{\frac{(r^2-a^2)(r-M)^2(r+3M)}{8(3M^5)^{1/4}(r^2+a^2)\Delta_K^{1/2}(r^{3/2}\pm aM^{1/2})}\right\}(4/\pi)(m/m_{crit})\,\varepsilon^{-3/2} + \right.$$

$$\left. + \; 2s\cdot\frac{(r+M)(r^2+a^2)}{\sqrt{3}\,m^2\omega_o^{5/2}\;r\Delta_K\varepsilon}\right]_{r=r_o} \ll 1 \qquad (8.43)$$

Since in the limit $r_o \simeq r_\gamma$ the factor in curly brackets has values $0 \leq \{\;\}_{r_o} \leq 3^{-1/2}$ for all values of $0 \leq a \leq M$, one may easily verify that (8.43) is satisfied for all multipoles m and spin s, as long $m_{crit} \gg 1$.

To find the barrier penetration factor ${}_s\Theta(r_o)$ the potential V_o is approximated by a parabola in the region of the peak $r = r_\gamma$

$$V_o(r) = V_o(r_\gamma) + \tfrac{1}{2}V_o''(r_\gamma)\;(r^*-r_\gamma^*)^2$$

$$\simeq \frac{r_\gamma^4\,W}{(r_\gamma^2+a^2)^2} + \frac{3\Delta_\gamma^2\,r_\gamma^2}{(r_\gamma^2+a^2)^4}\;(r^*-r_\gamma^*)^2 \;, \qquad (8.44)$$

and the spin-dependent term in ${}_s\Theta(r_o)$ is just a phase factor in the region where $V_o(r) > 0$. For $V_o(r) < 0$ it becomes a real factor giving the correct boundary conditions for ${}_su$ in the limit $r^* \to \pm\infty$. Inserting (8.44) into (8.35) and integrating from $r_o = (1+\delta)r_\gamma$ to the classical turning point $r_{tp} \simeq \left[1 + (W/3)^{1/2}\right]r_\gamma$ yields

$${}_s\Theta(r_o) \simeq \frac{\pi}{4}\varepsilon + \frac{is}{2}\int_{r_o^*}^{r_{tp}^*}(V_1/\kappa)\,dr^* \;, \qquad (8.45)$$

hence

$$\left|{}_su^{in}(r_o)\right|^2 \simeq \left[\frac{r^2+a^2}{\omega^{1/2}r_o^{1/2}\Delta_K^{1/2}3^{1/4}}\right]_{r=r_o}\varepsilon^{-1/2}\exp\left[-\pi\varepsilon/2\right]. \qquad (8.46)$$

The spin-dependence in (8.46) is of order ω^{-1} in the first factor and can therefore be neglected. Now we also take into account the normalization factors which were neglected in the derivation of (6.45). For the WKB solutions "in" and "out" (6.32) becomes [see Chrzanowski 1975, Eq. (5.6)]

$$ {}_sR^{out} = C\,\Delta_K^{-s}\,\overline{{}_{-s}R^{in}} \equiv \left(-\frac{i}{\omega}\right)^s 2^{s/2-1}\,\Delta_K^{-s}\,\overline{{}_{-s}R^{in}} , \qquad (8.47) $$

where ${}_sR^{out}$ satisfies the boundary conditions

$$ {}_sR^{out} \sim C \begin{cases} |\omega|^{-1/2}\left[\bar{S}\, r^{-1} e^{-i\omega r^*} + r^{-2s-1} e^{i\omega r^*}\right], & r^* \to \infty \\ |k_-|^{-1/2}\, \bar{T}\, e^{ik_- r^*} & , r^* \to -\infty . \end{cases} \qquad (8.58) $$

<u>High-frequency approximation for ${}_sZ^{\omega}_{\ell m}(\pi/2,0)$</u>

Since in this chapter only equatorial orbits are considered for the test particle motion it suffices to evaluate the spin-weighted harmonics and their derivatives at $\theta = \frac{\pi}{2}$. The idea (Chrzanowski & Misner 1974) is to linearize the angular equation about the equatorial plane and to reduce the resulting equation to a standard equation with known solutions. Again, it will be seen that for large frequencies spin has negligible influence on the equatorial values of the angular eigenfunctions. Consider the angular master equation (6.25), namely

$$ \left[\frac{d^2}{d\theta^2} + \cot\theta\,\frac{d}{d\theta} + a^2\omega^2\cos^2\theta - 2a\omega s\cos\theta - \frac{(m+s\cos\theta)^2}{\sin^2\theta} + {}_sA_{\ell m} + s\right]{}_sS_{\ell m} = 0. \qquad (8.49) $$

Performing the transformations

$$ \vartheta = \pi/2 - \theta, \qquad {}_sT_{\ell m} = \left[\cos\vartheta\right]^{-1/2}\,{}_sS_{\ell m} $$

changes Eq. (8.49) into

$$ \left[\frac{d^2}{d\vartheta^2} - \left(\frac{m^2+s^2-\frac{1}{4}}{\cos^2} - a^2\omega^2\right)\sin^2\vartheta + 2s\left(\frac{m}{\cos^2\vartheta} + a\omega\right) + {}_sQ + s + \frac{1}{2}\right]{}_sT_{\ell m} = 0, \qquad (8.50) $$

where ${}_sQ = {}_sA - m^2$. Eq. (8.50) may be solved as an effective potential problem for $m \gg 1$. The classical turning points are at $\vartheta_{tp} = \pm\, O(m^{-1/2})$.

Also, since the ${}_sT_{\ell m}$ are exponentially damped beyond ϑ_{tp} this equations can be linearized about $\vartheta = 0$. Setting $b = m/\omega$, then leads directly to

$$\left[\frac{d^2}{d\vartheta^2} - (1 - \frac{a^2}{b^2}) m^2 \vartheta^2 + 2s(1 + \frac{a}{b}) m\vartheta + {}_sQ \right] {}_sT_{\ell m} = 0, \qquad (8.51a)$$

where $s^2 - \frac{1}{2} \ll m^2$ and $s + \frac{1}{2} \ll {}_sQ \sim O(m)$ were neglected. Transforming Eq. (8.51a) to a new variable

$$\xi = m^{1/2} (1-a^2/b^2)^{1/4} \vartheta + s\, m^{-1/2} \beta^{1/4},$$

where $\beta \equiv (1+a/b)(1-a/b)^{-3}$, leads to

$$\left[\frac{d^2}{d\xi^2} + \frac{{}_sQ + s^2(1+a/b)(1-a/b)^{-1}}{m(1-a^2/b^2)^{1/2}} - \xi^2 \right] {}_sT_{\ell m} = 0. \qquad (8.51b)$$

By their definition the ${}_sT_{\ell m}$ vanish for $\vartheta \to \pm\pi/2$. Equivalently, in the linearized equation (8.51b) this boundary condition is replaced by the demand that the ${}_sT_{\ell m}$ vanish for $\xi \to \pm\infty$. Eq. (8.51b) together with these boundary conditions can be considered to be a harmonic oscillator problem. One identifies the eigenvalues ${}_sQ$ with

$${}_sQ = m(1-a^2/b^2)^{1/2}(2k+1) - \frac{1+a/b}{1-a/b} s^2 + O(m^0) \qquad (8.52)$$

and the eigenfunctions ${}_sT_{\ell m}$ with

$${}_sT_{\ell m} = {}_sC_{\ell m} H_k(\xi) \exp(-\xi^2/2) \qquad (8.53)$$

for some coefficients ${}_sC_{\ell m}$ and for $k = \ell - m$. Recall that the Hermite polynomials $H_k(\xi)$ satisfy

$$\frac{d}{d\xi} H_k(\xi) = 2k\, H_{k-1}(\xi) \quad ,$$

$$\int_{-\infty}^{\infty} d\xi \; e^{-\xi^2/2} \left[H_k(\xi)\right]^2 = 2^k \pi^{1/2} k! \quad ,$$

$$H_k(0) = \begin{cases} (-1)^{k/2}\, k!/(k/2)! & , k \text{ even}, \\ 0 & k \text{ odd}. \end{cases}$$

Using the last two relations and the normalization of the spin-weighted spheroidal harmonics as given by (6.23), the coefficients are determined to be

$$|{}_s C_{\ell m}|^2 = (1-a^2/b^2)^{1/4} \frac{m^{1/2}}{2^{k-1}\pi^{1/2}(k!)^2} \frac{(2m+k)!}{(2m+2k+1)!} . \qquad (8.54)$$

For $\vartheta = 0$ one has $\xi = s\, m^{-1/2} \beta^{1/4} \simeq 3sm^{-1/2} = O(m^{-1/2})$. Since the above approximation holds only up to order $O(m^0)$, one obtaines

$${}_s S_{\ell m}(-a\omega, \vartheta=0) = \begin{cases} (1-a^2/b^2)^{1/8} \dfrac{(-1)^{k/2}\sqrt{2}\, m^{1/4}}{\pi^{1/4}\, 2^{k/2} (\frac{k}{2})!} \left[\dfrac{(2m+k)!}{(2m+2k+1)!}\right]^{1/2}, & k \text{ even,} \\ 0 & , k \text{ odd .} \end{cases} \qquad (8.55)$$

Only the first factor in (8.54) depends on the Kerr parameter a. In addition, note that for $a = 0$, ${}_s Z^{\omega}_{\ell m}(\pi/2,0) \cong {}_s Y_{\ell m}(\pi/2,0)$. Therefore, for even k one may write

$$\begin{aligned} |{}_s Z^{\omega}_{\ell m}(\pi/2,0)|^2 &= (1-a^2/b^2)^{1/4} |{}_s Y_{\ell m}(\pi/2,0)|^2 \\ &= (1-a^2/b^2)^{1/4} \frac{k!}{2\pi^{3/2}(\frac{k}{2}!)^2 2^k} m^{1/2} . \end{aligned} \qquad (8.56)$$

The terms for odd k are of lower order in m and will therefore be neglected. (For a more precise statement of this equation see Appendix B.) The derivative $\frac{d}{d\vartheta}\, {}_s Z^{\omega}_{\ell m}(\pi/2,0)$ may be found with the aid of

$$\frac{d}{d\vartheta}\, {}_s T_{\ell m}\Big|_{\vartheta=0} = \frac{d\xi}{d\vartheta}\frac{d}{d\xi}\, {}_s T_{\ell m}\Big|_{\vartheta=0} \qquad (8.57)$$

together with the previously cited relation for the derivative of Hermite polynomials. Thus

$$\begin{aligned} \frac{d}{d\vartheta}\, {}_s T_{\ell m}\Big|_{\vartheta=0} &= m^{1/2}(1-a^2/b^2)^{1/4}\, {}_s C_{mk}\, 2k\, H_{k-1}(0) \frac{{}_s C_{m+1,k-1}}{{}_s C_{m+1,k-1}} \\ &= 2k\, m^{1/2}(1-a^2/b^2)^{1/4} \frac{{}_s C_{mk}}{{}_s C_{m+1,k-1}}\, {}_s T_{\ell,m-1}(\vartheta=0) . \end{aligned}$$

The ratio of the coefficients is given by

$$\frac{{}_sC_{mk}}{{}_sC_{m+1,k-1}} = \left[\sqrt{2}\; k \;(2m+k+1)^{1/2}\right]^{-1/2}.$$

Therefore,

$$\left|\frac{d}{d\theta}\, {}_sZ^{\omega}_{\ell m}(\pi/2,0)\right|^2 = 2km(1-a^2/b^2)^{1/2}\left|\,{}_sZ^{\omega}_{\ell,m+1}(\pi/2,0)\right|^2 \tag{8.58}$$

This result can be obtained more directly from Eq. (8.48). The quantity $E = {}_sQ$ as given by (8.52) may be identified with the "energy" of the (ℓ,m) mode in the equatorial plane. Using the WKB approximation together the asymmetry of spin-weighted spheroidal harmonics with resprect to the equatorial plane one concludes that

$$\begin{aligned}\frac{d}{d\theta}\, {}_sZ^{\omega}_{\ell m} &= {}_sQ^{1/2}\; {}_sZ^{\omega}_{\ell,m+1} \\ &= \left[m(1-a^2/b^2)^{1/2}(2k+1) - \frac{1+a/b}{1-a/b}\, s^2\right]^{1/2} {}_sZ^{\omega}_{\ell,m+1}.\end{aligned} \tag{8.59}$$

This agrees with (8.58) to leading order in m. A similar kind of argument leads to the expression

$$\begin{aligned}\left|\frac{d^2}{d\theta^2}\, {}_sZ^{\omega}_{\ell m}(\pi/2,0)\right|^2 &= {}_sQ^2\left|{}_sZ^{\omega}_{\ell m}(\pi/2,0)\right|^2 \\ &\simeq m^2\varepsilon^2(1-a^2\omega_o^2)\left|\,{}_sZ^{\omega}_{\ell m}(\pi/2,0)\right|^2\end{aligned} \tag{8.60}$$

for the second derivative of ${}_sZ^{\omega}_{\ell m}$.

8.4 Scalar GSR in the Kerr Geometry

The solution of the scalar wave equation (6.20) with $s = 0$ can according to Eq. (6.54) be written as

$$\phi(x) = \oint i\, \frac{\omega}{|\omega|}\, \phi^{up}(x,\ell m\omega) \langle \phi^{out}(x,\ell m\omega),\, fT\rangle . \tag{8.61}$$

Here ϕ^{up} and ϕ^{out} as defined by (6.37) are solutions of the homogeneous wave equation that satisfy the boundary conditions (6.38). The solutions $\phi(x,\ell m\omega)$ are normalized such that the condition

$$\langle \phi(x,\ell m\omega),\ \phi(x,\ell' m'\omega') \rangle = \frac{\omega}{|\omega|}\, \delta_{\ell\ell'}\, \delta_{mm'}\, \delta(\omega-\omega')$$

holds at least on $\mathcal{J}^+$, where the completeness of the expansion (8.61) can be shown with the aid of the asymptotic properties of ϕ. The two fields appearing in $\langle,\rangle$ are either (up, out) or (in,down).

The frequency spectrum of radiation reaching $\mathcal{J}^+$ is derived from

$$E^{(o)}_{out} = -\lim_{r\to\infty} \int d\Omega\, \Sigma T^r{}_t = \lim_{r\to\infty} \int d\Omega dt\, \frac{r^2\omega^2}{4\pi}\, |\phi|^2 \equiv \int_{\omega>0} d\omega\, \frac{dE^{(o)}_{out}}{d\omega}, \qquad (8.62)$$

using (8.3). Then substituting (8.61) into (8.62) it follows that

$$\frac{dE^{(o)}_{out}}{d\omega} = \sum_{\ell=0}^{\infty} \sum_{m=-\ell}^{+\ell} \omega \left| \langle \phi^{out}(x,\ell m\omega), fT \rangle \right|^2 . \qquad (8.63)$$

The energy flow down the black hole is described by

$$E^{(o)}_{down} = +\lim_{r^*\to-\infty} \int d\Omega\, \Sigma T^r{}_t = \lim_{r^*\to-\infty} \int d\Omega\, \Sigma \frac{(\omega-m\omega_h)\omega}{4\pi}\, |\phi|^2 .$$

Hence $E^{(o)}_{down} < 0$ whenever $\omega < m\omega_h$ which means that energy is extracted from the black hole by ingoing waves carrying negative energy.

This phenomenon - first noted by Zel'dovich (1971, 1972) and indepentently by Misner (1972) - is the wave analogue of the Penrose mechanism, where particles can break up into (at least) two pieces within the ergosphere in a manner, so that one goes down the black hole in a (necessarily counterrotating) negative-energy-at-infinity orbit, while the other piece can escape to infinity carrying more than the total energy of the initial part. The energy extracted is rotational energy of the black hole. This process is impossible for a nonrotating black hole; there, the limiting surfaces of the ergosphere - the horizon and the infinite-redshift-surface - coincide.

Analogously, waves can bounce back from the black hole with amplified energy. The "amount of interest" (Misner) the black hole is paying in this process is always less than one percent in the scalar case (Press and Teukolsky 1972) and up to $\sim$ 140 % for gravitational radiation (Starobinsky 1973, Starobinsky & Churilov 1973, Press & Teukolsky 1974). If the wave could bounce back many times to the black hole from some appropriate impenetrable mirror enclosing the black hole, the wave could be enhanced considerably. This then would lead, at the final eruption where the mirror is destroyed, to a so-called black hole bomb (Press & Teukolsky 1972).

Another particular consequence is that the particles radiating negative energy waves down the black hole gain orbital energy and increase their orbit radius on account of the black hole's angular momentum, until equilibrium is re-established, i.e. until $\omega - m\omega_h$ becomes zero or positive again ("floating orbits"). For circular, equatorial particle motion the scalar source term becomes

$$\sqrt{-g}\, T = -\mu \left(\frac{dt}{d\tau}\right)^{-1} \delta(r-r_o)\, \delta(\theta-\pi/2)\, \delta(\varphi-\omega_o t) \ , \qquad (8.64)$$

where $\omega_o = \frac{d\varphi}{dt}$ and $dt/d\tau$ were calculated in Sec. 3.3. Evaluation of (8.63) then leads to the power spectrum

$$\frac{dP^{(o)}_{out}}{d\omega} = \sum_{\ell \geq |m|} \frac{2\pi f^2 \mu^2 m\, \omega_o}{(dt/d\tau)^2} \frac{\left|{}_o u^{out}(r_o^*)\right|^2}{r_o^2 + a^2} \left|{}_o Z^{\omega}_{\ell m}(\pi/2,0)\right|^2 , \qquad (8.65)$$

The radial and angular functions appearing in this equation were calculated in the preceding section. Inserting (8.46) and (8.56) into (8.65) one obtains for the mode (k,m) (k even) the power

$$P^{(o)}_{out}(k,m) = \frac{8 f^2 \mu^2 \sqrt{3M}}{\pi^{3/2} r_\gamma^{3/2} (r_\gamma + 3M)^2} (r_\gamma - M) \frac{k!}{(\frac{k}{2}!)^2 2^k} \left(\frac{m}{m_{crit}}\right) \varepsilon^{-1/2} e^{-\pi\varepsilon/2} . \qquad (8.66)$$

For the total power (8.66) has to be summed over all k and m, i.e.

$$P^{(o)}_{out} = \sum_{m=0} \sum_{\substack{k=0 \\ \text{even}}} P^{(o)}_{out}(k,m) = \sum_{m=0}^{\infty} \frac{dP^{(o)}_{out}}{d\omega} \Delta\omega \simeq \int_0^{\infty} dm\, \omega_o \frac{dP^{(o)}_{out}}{d\omega}$$

where $$\frac{dP^{(o)}_{out}}{d\omega} = \frac{1}{\omega_o} \sum_{\substack{k=0\\ \text{even}}}^{\infty} P^{(o)}_{out}(k,m) \; .$$

As in Sec. 8.2, the summation over k can be truncated at $k = k_{crit} = 0$; this gives

$$\frac{dP^{(o)}_{out}}{d\omega} = \frac{4f^2\mu^2\sqrt{3}}{\pi^{3/2}r_\gamma} \frac{(r_\gamma - M)}{(r_\gamma + 3M)} \quad \varepsilon^{-1/2} \left| \frac{\omega}{\omega_{crit}} \right| \exp(-\pi\varepsilon/2) , \tag{8.67}$$

where

$$\omega_{crit} = \frac{2}{\pi}\sqrt{3M} \frac{r_\gamma + 3M}{\sqrt{r_\gamma M}\,(r_\gamma^{3/2} + aM^{1/2})} = \frac{4\sqrt{3}\,M^{1/2}}{\pi r_\gamma (r_\gamma + 3M)} \; . \tag{8.68}$$

The angle of semi-width of the angular power distribution may be seen from (8.60) to be

$$\Delta\vartheta = |m|^{-1/2} \left[1 - a^2\omega_o^2\right]^{-1/4} . \tag{6.69}$$

8.5 Electromagnetic GSR in the Kerr Geometry

Chrzanowski & Misner (1974) have argued that both the electromagnetic and gravitational solutions of the wave equation can be factorized in a fashion analogous to the factorization of the scalar potential in (8.61). The form of the electromagnetic potential should be

$$A_a(x) = \sum\!\!\!\!\!\!\int \; i \frac{\omega}{|\omega|} A^{up}_a(x,\ell m\omega P) \left\langle A^{out}_a(x,\ell m\omega P) , J^a \right\rangle , \tag{8.70}$$

where the summation now includes also a sum over the two polarization states P defined in Eq. (6.55 & 58). The existence of the form (8.70) could not be proved in general. However, plausibility arguments have been given that it should hold at least on $\mathscr{I}^+$ together with the normalization condition

$$\left\langle A_a(s,\ell m\omega P), A^a(x,\ell' m'\omega' P')\right\rangle_{\mathscr{I}^+} = -\frac{\omega}{|\omega|}\,\delta_{PP'}\,\delta_{\ell\ell'}\,\delta_{mm'}\,\delta(\omega - \omega') . \tag{8.71}$$

Again, the difficult part is the proof of completeness of the decomposition (8,70), a problem discussed in Sec. 6.9. The electromagnetic energy radiation to $\mathcal{J}^+$ is defined by

$$E^{(1)}_{out} = - \lim_{r \to \infty} \int d\Omega r^2 T^{(1)r}_{\ t} = \lim_{r \to \infty} \int d\Omega dt \frac{r^2}{2\pi} |\phi_2|^2, \tag{8.72}$$

where for the last expression the energy-momentum tensor of the electromagnetic field has been used in the form

$$\begin{aligned} 4\pi T^{(1)}_{ab} = \{ |\phi_o|^2 n_a n_b + 2|\phi_1|^2 [\ell_{(a} n_{b)} + m_{(a} \bar{m}_{b)}] + |\phi_2|^2 \ell_a \ell_b - \\ - 4 \bar{\phi}_o \phi_1 n_{(a} m_{b)} - 4 \bar{\phi}_1 \phi_2 \ell_{(a} m_{b)} + 2 \bar{\phi}_o \phi_2 m_a m_b \} + c.c. \end{aligned} \tag{8.73}$$

Since for $r \to \infty$ the leading terms are given by

$$T^{(1)r}_{\ t} \sim \frac{1}{4\pi} (A^{\theta;r} A_{\theta;t} + A^{\varphi;r} A_{\varphi;t})$$

one concludes, analogous to (8.63), that the spectrum will be

$$\frac{dE^{(1)}_{out}}{d\omega} = \sum_{\ell > |m|} \sum_{P=\pm} \omega |\langle A^{out}_a (x,\ell m\omega P), J^a \rangle|^2 . \tag{8.74}$$

The sum over P includes the two transverse polarization states in the θ-and φ-direction.

Sources

The current-vector of a test charge q at a point $z^a(\tau)$ and four-velocity u^a, as defined in (8.2), is given by

$$J^a = \frac{q}{\sqrt{-g}} \frac{u^a}{u^o} \delta(r-R) \delta(\theta - \Theta) \delta(\varphi - \Phi). \tag{8.75}$$

Its tetrad components are $J_\ell = J_a \ell^a$, $J_n = J_a n^a$ and $J_m = J_a m^a$, respectively. When expressed in tetrad (6.12a), the tetrad components of u^a required here are

$$u^a \ell_a = \frac{u^o}{\Delta_K} \left\{ (r^2+a^2)(1- \frac{2Mr}{\Sigma} + \frac{N}{\Sigma}\dot{\Phi}) - \Sigma \dot{R} + \frac{aN}{\Sigma} - a \sin^2\theta (r^2+a^2+ \frac{aN}{\Sigma}) \dot{\Phi} \right\},$$

$$u^a n_a = \frac{u^o}{2\Sigma}\left\{(r^2+a^2)(1-\frac{2Mr}{\Sigma}+\frac{N}{\Sigma}\dot{\Phi}) + \Sigma\dot{R} - a\sin^2\theta\,(r^2+a^2+\frac{aN}{\Sigma})\dot{\Phi}\right\},$$

$$u^a m_a = \frac{u^o\bar{\varsigma}}{\sqrt{2}}\left\{(1-\frac{2Mr}{\Sigma}+\frac{N}{\Sigma}\dot{\Phi})\,ia\sin\theta - \Sigma\dot{\Theta} - \frac{i}{\sin\theta}\left[\sin^2\theta\,(r^2+a^2+\frac{Na}{\Sigma})\dot{\Phi} - \frac{N}{\Sigma}\right]\right\}, \tag{8.76a}$$

where $N \equiv 2\,Mar\sin^2\theta$. For circular equatorial orbits

$$z^a = (t, r_o, \pi/2, \omega_o t),\quad u^a = u^o(1,0,0,\omega_o) \text{ and}$$

$$u^a \ell_a = u^o(1-a\omega_o) = u^o\,\frac{2r_\gamma}{r_\gamma+3M},\quad u^a n_a = \frac{u^o}{2r_o^2}\Delta_K(1-a\omega_o) = \frac{u^o(r_\gamma-M)^2}{4M(r_\gamma+3M)},$$

$$u^a m_a = -\frac{iu^o}{\sqrt{2}\,r_o}\left\{\omega_o(r_o^2+a^2) - a\right\} = -\,iu^o\,\frac{r_\gamma}{\sqrt{6M}}\,\frac{(r_\gamma-M)}{(r_\gamma+3M)}, \tag{8.76b}$$

where (3.26a) was used to find the expressions involving r_γ .
The inner product appearing in (8.74) can be expressed with the aid of (6.57) as

$$\langle A_a^{out}, J^a\rangle = \int d^4x_o\sqrt{-g}\left[{}_{-1}\bar{X}^{out} J_n + {}_{-1}\bar{X}_m^{out} J_m \pm ({}_{-1}X^{out} J_n + {}_{-1}X^{out}\bar{J}_m)\right]$$

$$= \frac{q}{u^o}\left[u^a n_a A_1 + u^a m_a A_2 \pm (u^a n_a A_3 + u^a\bar{m}_a A_4)\right]$$

$$= \frac{q}{u^o}\left[u^a n_a(A_1 \pm A_3) + u^a m_a(A_2 \mp A_4)\right], \tag{8.77}$$

since $(u^a\bar{m}_a) = -(u^a m_a)$ for circular orbits. For this case the obvious abbreviations $A_1,\ldots,A_4$ become

$$A_1 = \frac{r_o}{\sqrt{2}}\left\{\left[m(a\omega_o-1) - \frac{ia}{r_o}\right]{}_{-1}\bar{Z} + \partial_\theta\,{}_{-1}\bar{Z}\right\}{}_{+1}\bar{R}^{out}e^{i\omega t}$$

$$A_2 = \frac{1}{2}\left\{\left[im\left(\omega_o(r_o^2+a^2) - a\right) - r_o + \frac{a^2}{r_o}\right]{}_{+1}\bar{R}^{out} - (r_o^2+a^2)\frac{d}{dr^*}\,{}_{+1}\bar{R}^{out}\right\}{}_{-1}\bar{Z}e^{i\omega t},$$

$$A_3 = \frac{r_o}{\sqrt{2}}\left\{\left[-m(a\omega_o-1) + \frac{ia}{r_o}\right]{}_{+1}\bar{Z} + \partial_\theta\,{}_{+1}\bar{Z}\right\}{}_{+1}\bar{R}^{out}e^{i\omega t}, \tag{8.78}$$

$$A_4 = \frac{1}{2}\left\{\left[im\left(\omega_o(r_o^2+a^2)-a\right) - r_o + \frac{a^2}{r_o}\right]{}_{+1}\bar{R}^{out} - (r_o^2+a^2)\frac{d}{dr^*}\,{}_{+1}\bar{R}^{out}\right\}{}_{+1}\bar{Z}e^{i\omega t},$$

where $${}_{\pm1}R^{out}\,{}_{\pm1}Z \equiv {}_{\pm1}R^{out}(r_o^*)\,{}_{\pm1}Z(\pi/2,0) .$$

In order to simplify (8.77) one uses from (6.32)

$$ {}_{+s}\bar{Z}(\pi/2,0) = (-1)^{\ell+m} {}_{-s}\bar{Z}(\pi/2,0), \qquad {}_{+s}\bar{Z}(\pi/2,0) = -(-1)^{\ell+m} {}_{-s}\bar{Z}(\pi/2,0), \tag{8.79} $$

and therefore (taking the leading terms in m)

$$ A_3 = (-1)^{\ell+m+1} A_1 \quad , \qquad A_4 = (-1)^{\ell+m} A_2 . $$

Hence

$$ \langle A_a^{out}, J^a \rangle = \frac{q}{u^o} \left[1 \pm (-1)^{\ell+m+1} \right] \left[(u^a n_a) A_1 + (u^a m_a) A_2 \right. $$

$$ = \frac{2q}{u^o} \left[(u^a n_a) A_1 + (u^a m_a) A_2 \right] . \tag{8.80} $$

since for even $(\ell+m+1)$ only the (+) state contributes and for odd $(\ell+m+1)$ only the (-) state is nonzero, i.e. there is only one polarization state present. Inserting (8.80) into Eq. (8.74) yields for the power per unit frequency

$$ \frac{dP^{(1)}_{out}}{d\omega} \simeq A^{(1)} \, \varepsilon^{-1/2} \exp(-\pi\varepsilon/2) \quad , \tag{8.81} $$

where

$$ A^{(1)} = \left(\frac{q}{M}\right)^2 \frac{M^{1/2} r_\gamma^4}{18\pi^{1/2} r_\gamma^{1/2} \, M^4} (r_\gamma - M) \tag{8.82} $$

In Eq. (8.81) only the even parity term appears in the power formula since the odd parity component was neglected in the high-frequency approximation for the angular functions ${}_sZ^{\omega}_{\ell m}$ in (8.56). The contribution of the odd parity terms to the radiated power is less than 4 %, hence the radiation is ~ 96 % linearly polarized. This agrees with the computation of electromagnetic GSR in the Schwarzschild geometry by Breuer et al. (1973) and Breuer & Vishveshwara (1973) using Regge-Wheeler formalism and the separated equations of Ruffini et al. (1972). In fact, it was shown that at all latitudes relative to the orbital plane within the half-width of the radiation beam the degree of polarization is greater than 90 %. In Fig. 16 the electromagnetic spectrum as a function of both latitude and frequency is shown for GSR in the Schwarzschild geometry

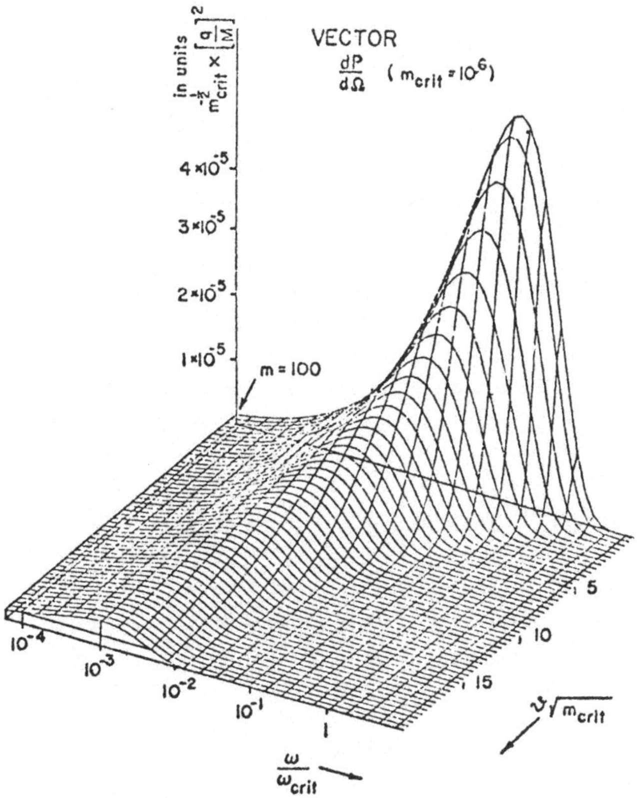

Figure 16. Electromagnetic power of GSR per unit solid angle and frequency in the Schwarzschild geometry (from Breuer & Vishveshwara 1973). The latitude-angle $\vartheta = \pi/2 - \theta$ (θ = polar angle) relative to the plane of the orbit is chosen such that the half-width of the beam is at $\vartheta\sqrt{m_{crit}} = 1$ for $\omega/\omega_{crit} = 1$.

8.6 Gravitational GSR in the Kerr geometry

The solution of the gravitational wave equation (6.20) with s=-2, according to Eq. (6.54), can be written as

$$h_{ab}(x) = \oint 2i \frac{\omega}{|\omega|} h^{up}_{ab} (\ell m \omega P x) . \langle h^{out}_{ab}(\ell m \omega P) \ , \ T^{ab} \rangle . \qquad (8.83)$$

The "up" and "out" solutions have a common asymptotic form at $\mathcal{J}^+$ (see Fig. 9) and are normalized by the condition

$$\langle h_{ab}(\ell m \omega P), \ h^{ab}(\ell' m' \omega' P') \rangle = - \frac{2\omega}{|\omega|} \delta_{PP'} \delta_{\ell\ell'} \delta_{mm'} \delta(\omega - \omega') . \qquad (8.84)$$

The gravitational energy flux radiated to $\mathcal{J}^+$ is (cf. MTW)

$$E^{(2)}_{out} = -\lim_{r\to\infty} \int d\Omega \, dt \; r^2 T^r{}_t = \lim_{r\to\infty} \int d\Omega \, dt \; \frac{r^2}{4\pi \omega^2} |\Psi_4|^2 . \qquad (8.85)$$

where T_{ab} is the Isaacson effective energy-momentum tensor. Substituting (8.93) into (8.95) one obtains the energy flux in the form

$$E^{(2)}_{out} = \oint_{\omega>0} \omega |\langle h^{out}_{ab}(\ell m m P), \ T^{ab} \rangle|^2 \equiv \int_0^\infty d\omega \frac{dE^{(2)}_{out}}{d\omega} . \qquad (8.86)$$

The energy momentum tensor corresponding to a testparticle of mass μ and 4-velocity u^a is given by

$$T^{ab} = \mu (-g)^{-1/2} (u^o)^{-1} \; u^a u^b \; \delta(r-R) \;\; \delta(\theta - \Theta) \, \delta(\varphi - \Phi) . \qquad (8.87)$$

Its tetrad components are $T_{\ell\ell} = T_{ab} \ell^a \ell^b$, $T_{\ell n} = T_{ab} \ell^a n^b$ etc. In Eq. (8.86) the inner product $\langle , \rangle$ can be written explicitely using Eq. (6.57) (in the trace-free, outgoing gauge, $h_{ab} n^b = h_a{}^a = 0$) as

$$\begin{aligned}
\langle h^{out}_{ab}, T^{ab} \rangle = \int d^4x_o \sqrt{-g} \Big[{}_{-2}\bar{X}^{out} \; T_{nn} - {}_{-2}\bar{X}^{out}_{(\ell\bar{m})} T_{n\bar{m}} + {}_{-2}\bar{X}^{out}_{\bar{m}\bar{m}} \; T_{\bar{m}\bar{m}} \pm \\
\pm ({}_{-2}X^{out} \; T_{nn} - {}_{-2}X^{out}_{(\ell m)} \; \bar{T}_{nm} + {}_{-2}X^{out}_{mm} \; \bar{T}_{mm}) \Big]
\end{aligned}$$

$$\begin{aligned}
= \frac{\mu}{u^o} \Big[\; (u^a n_a)^2 \; H_1 \; - \; (u^a n_a)(u^a \bar{m}_a) \; H_2 \; + \; (u^a \bar{m}_a)^2 \; H_3 \quad \pm \\
\pm \Big\{ (u^a n_a)^2 \; H_4 \; - \; (u^a n_a)(u^a m_a) \; H_5 \; + \; (u^a m_a) \; H_6 \Big\} \Big] . \qquad (8.88)
\end{aligned}$$

For circular orbits the quantities H_i become up to order m

$$H_1 = \frac{1}{2\varrho^4}\left\{ \varrho\left[a\omega\sin\theta - \frac{m}{\sin\theta} - 2ia\varrho\sin\theta - \cot\theta + \partial_\theta\right]\left[\varrho(a\omega\sin\theta - \right.\right.$$

$$\left.\left. - \frac{m}{\sin\theta} + 3ia\varrho\sin\theta + 2\cot\theta + \partial_\theta)\right]\right\}{}_{+2}\bar{R}^{out}\,{}_{-2}\bar{Z}\; e^{i\omega t}$$

$$\simeq \frac{r_o^{\,2}}{2}\; m^2(1-a\omega_o)^2\; {}_{+2}\bar{R}^{out}\,{}_{+2}\bar{Z}\; e^{i\omega t} \quad ,$$

$$H_2 = \frac{1}{4\varrho^4\Sigma}\left[i\omega(r^2+a^2) - ima + \Delta_K(2\varrho+\bar{\varrho}) - 2(r-M) - \Delta_K\,\partial_r\right]\times$$

$$\times\left\{\Sigma^{-1}\left[i\omega(r^2+a^2) - ima - 3\varrho\Delta_K - 4(r-M) - (r^2+a^2)\,\partial_{r*}\right]\right\}{}_{+2}\bar{R}^{out}\,{}_{-2}\bar{Z}\; e^{i\omega t},$$

$$\simeq -\frac{1}{4}\; m^2\left[\omega_o(r_o^{\,2}+a^2) - a\right]^2\; {}_{+2}\bar{R}^{out}\,{}_{-2}\bar{Z}\; e^{i\omega t}, \qquad (8.89)$$

$$H_3 = -(2\sqrt{2}\varrho^4)^{-1}\left\{ \varrho\,(a\omega\sin\theta - \frac{m}{\sin\theta} + 2\cot\theta - ia\sin\theta(\varrho+2\bar{\varrho}) + \partial_\theta)\times\right.$$

$$\times\left[\Sigma^{-1}(a\omega\sin\theta - \frac{m}{\sin\theta} - 3\varrho\Delta_K - 4(r-M) - (r^2+a^2)\,\partial_{r*})\right] +$$

$$+\Sigma^{-1}\left(i\omega(r^2+a^2) - iam + 4\Delta_K(\varrho-\bar{\varrho}) - 4(r-M) - \Delta_K\,\partial_r\right)\left[\varrho\,(a\omega\sin\theta - \right.$$

$$\left.\left. - \frac{m}{\sin\theta} + 3ia\varrho\sin\theta + 2\cot\theta + \partial_\theta\right]\right\}\; {}_{+2}\bar{R}^{out}\,{}_{-2}\bar{Z}\; e^{i\omega t}$$

$$\simeq -\frac{r_o}{\sqrt{2}}\; im^2(1-a\omega_o)\left[\omega_o(r_o^{\,2}+a^2) - a\right]{}_{+2}\bar{R}^{out}\,{}_{-2}\bar{Z}\; e^{i\omega t} \;.$$

Similar expressions result for H_4, H_5 & H_6, but using (8.79) one notes that for the leading terms

$$H_4 = (-1)^{\ell+m+1}H_1\;, \qquad H_5 = (-1)^{\ell+m}H_2, \qquad H_6 = (-1)^{\ell+m+1}H_3\;.$$

Therefore Eq. (8.88) becomes

$$\langle h^{out}_{ab},\, T^{ab}\rangle = \frac{2\mu}{u^o}\left[(u^a n_a)^2\, H_1 + (u^a n_a)(u^a m_a)H_2 + (u^a m_a)^2\, H_3\right]. \qquad (8.90)$$

Again, only one polarization state is present in the equatorial plane. With the above relations and the relativistic circular orbit equations the spectrum (8.86) is given by

$$\frac{dP^{(2)}_{out}}{d\omega} = \left(\frac{\mu}{M}\right)^2 \frac{\pi^{1/2}}{8\cdot 27}\frac{r_\gamma^8}{M^8}\frac{M^{1/2}(r_\gamma+3M)(r_\gamma-M)}{\epsilon^{1/2}}\left|\frac{\omega_{crit}}{\omega}\right| \exp(-\pi\epsilon/2) \; . \qquad (8.91)$$

Like in the electromagnetic case, the odd parity modes are absent in (8.91) since they have been neglected in (8.56). The Schwarzschild limit (a=0) of the power formula (8.91)(cf. Breuer et al. 1975) is plotted in Fig. 17 for different values of energy γ^2. The front factor in (8.91) as a function of the Kerr parameter a is discussed in Sec.8.1.

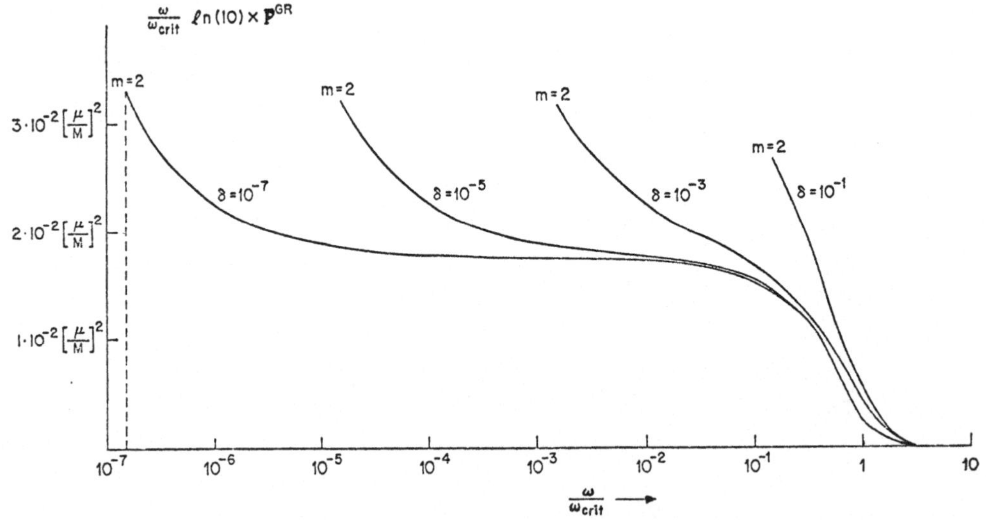

Figure 17. Power per unit frequency for gravitational GSR in the Schwarzschild geometry plotted for different values of the energy $\gamma^2 = (3\delta)^{-1}$. The closer the particle moves to the circular photon orbit the higher the multipoles $m=\omega/\omega_0$ are excited. The spectra expand to the left since $\omega_{crit}=4\omega_0/(\pi\delta)$ is held fixed. If plotted against m (i.e. m=2 held fixed) the cutoff would shift to the right with increasing particle energy.

IX. DISCUSSION

In this concluding chapter the GSR spectra for scalar, electromagnetic and gravitational radiation are discussed together with the conceptual limitations underlying the particular computations of high-frequency waves emitted by relativistic geodesic test particles. Furthermore, several open problems which occurred during the outline of gravitational perturbations theory are listed in the last section in order to show to the reader that radiation theory still faces a number of fundamental questions.

9.1 Master Formula for Power Spectra

The main result of Chap. XIII, the analysis of high-frequency radiation from test particles in unstable relativistic orbits about a Kerr black hole is summarized by the following master formula for the power spectrum

$$\frac{dP^{(s)}_{out}}{d\omega} \sim \left(\frac{f_s \mu}{M}\right)^2 A^{(s)}(r_\gamma) \left(\frac{\omega}{\omega_{crit}}\right)^{1-s} \exp\left[-2\omega/\omega_{crit}\right] , \qquad (9.1)$$

where $f_s = (f, q/\mu, 1)$. The frequency-independent amplitude $A^{(s)}(r_\gamma)$, the cutoff frequency ω_{crit} and the beaming angle $\Delta\vartheta$ are given by

$$\omega_{crit} = m_{crit}\,\omega_o = \frac{4\sqrt{3}}{\pi\, r_\gamma}\,\gamma^2, \qquad \Delta\vartheta = |m|^{-1/2}\left[1-a^2\omega_o^{\,2}\right]^{-1/4}, \qquad (9.2)$$

$$A^{(s)}(r_\gamma) = \frac{r^{4s-2}\,(r_\gamma+3M)^{1-s}\,(3Mr_\gamma)^{-s}}{\pi^{2-s}\,M^{5s}}\,A(a) ,$$

and

$$A(a) = \frac{2(r_\gamma - M)(3Mr_\gamma)^{1/2} M^2}{\pi^{1/2}(r_\gamma + 3M)^2 r_\gamma^2} . \tag{9.4}$$

These functions are plotted in Fig. 18 and the comparison of the different GSR spectra is given in Fig. 19. In (9.1) and subsequently s denotes the spin of the field under consideration (s=0, scalar; s=1, electromagnetic; s=2, gravitational). Eq. (9.1) follows directly from Eqs. (8.67), (8.81) and (8.91). From the master formula it follows that at high frequencies $(\omega \gg \omega_{crit})$, all spectra show qualitatively the behaviour characteristic of massless fields, while at low frequencies $(\omega \lesssim \omega_{crit})$, they differ by a factor ω^{1-s} (see Fig. 19). Eq. (9.1) and Fig. 18 also show that the rotation of the black hole does not alter the shape of the spectra; rather it serves as an overall amplification of the power. The half-width of the radiation beam given in (9.2) is only weakly dependent on a/M; in fact, it widens slightly for $a \to M$.

The radiation is 100% linearly polarized in the plane of the orbit. In view of the original attempt to explain Weber's observations as GSR - unrealistic because of the instability of relativistic geodesics - it was also the high degree of polarization which conflicted with an upper limit of ~ 40 % polarization set by Tyson & Douglass (1972) resulting from an analysis of Weber's data. Hence depolarization mechanisms were sought. Hughes & Misner (unpublished) found no significant rotation of polarization axes, no "gravitational Faraday rotations" for radiation near the equatorial plane of the Kerr metric. Gerlach (1974) however pointed out that the polarization of suitably excited electromagnetic and gravitational waves in the vicinity of charged black holes undergo significant Faraday rotation. In addition, the two waves undergo an interconversion-process of energy in the presence of a background electromagnetic field.

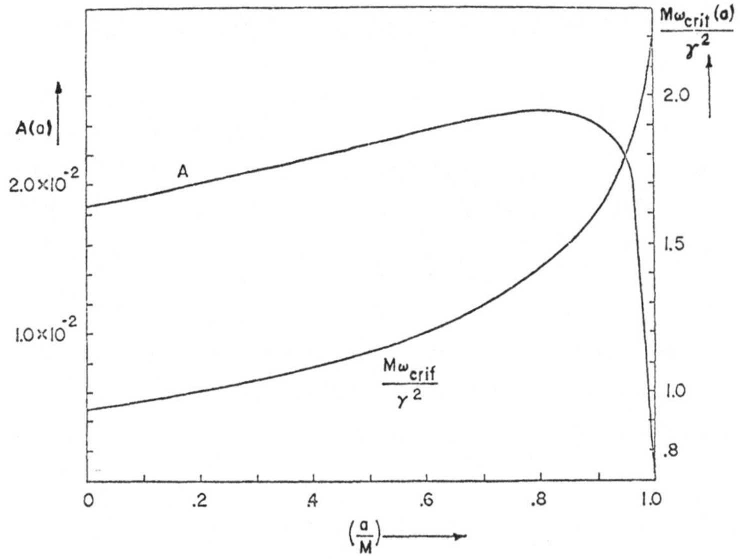

Figure 18. Cutoff frequency and amplitude factor of GSR spectra as a function of the Kerr parameter a/M.

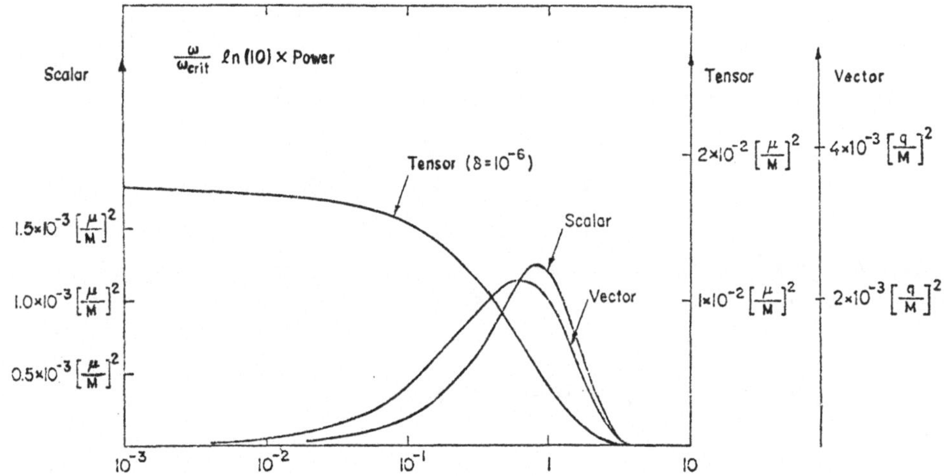

Figure 19. Comparison of scalar, electromagnetic & gravitational GSR spectra in the Schwarzschild geometry (from Breuer et al. 1973). The quantitiy $(\omega/\omega_{crit})\,\ell n(10)\times$Power is plotted, so the area under these curves represents the total energy when plotted against $\log_{10}(\omega/\omega_{crit})$. The gravitational extends to the left till m=2 or $2\omega_o/\omega_{crit} = \pi\delta/2 = 10^{-5.8}$ for $\delta = 10^{-6}$, where δ is defined by $r_o = r_\gamma\,(1+\delta)$.

Cutoff frequency and angular spreading of radiation pulses

All GSR power spectra have a characteristic exponential cutoff at frequencies higher than $\omega_{crit} \sim \gamma^2/M$, a result which should be compared to $\omega_{crit} \sim \gamma^3/r_o$ for synchrotron radiation in flat spacetime (Chap. IV). This difference is due to the different nature of synchrotron emission of photons depending on whether or not the source is accelerated (cf. Chrzanowski & Misner 1974).

Since radiation from a relativistic particle (in flat spacetime) is forwardly beamed within a cone of half-width γ^{-1}, it is clear from Fig. 20a that radiation will only reach an observer who lies within an angle $\Delta\varphi \sim \gamma^{-1}$ of the forward direction of the moving particle. Thus the observer receives radiation from circular orbits as searchlight pulses with azimuthal width $\Delta\varphi \sim \gamma^{-1}$. The first and last part of such a pulse reach a particular observer at times t_1 & t_2, respectively, where referring to Fig. 20a

$$t_1 = (t_{AB})_{rad} + (t_{B\infty})_{rad}, \qquad t_2 = (t_{AB})_{part} + (t_{B\infty})_{rad}. \tag{9.5}$$

Hence the observer receives radiation for a time

$$\Delta t = t_2 - t_1 \sim r_o \Delta\varphi \ (v^{-1}-1) \sim \frac{r_o}{2} \gamma^{-3} , \tag{9.6}$$

where $\Delta\varphi \sim \gamma^{-1}$ has been used. From Fourier transforms one concludes that there is only significant radiation at frequencies

$$\omega \lesssim \omega_{crit} \sim (\Delta t)^{-1} \sim \gamma^3/r_o . \tag{9.7}$$

Assuming for GSR $\omega_{crit} \sim \gamma^2/M$, it follows from (9.7) that an observer receives pulses of duration $\Delta t \sim \omega_{crit}^{-1} \sim M/\gamma^2$. Radiation reaches an observer when emitted between times (see Fig. 20b)

$$t_1 = (t_{OB})_{rad} + (t_{B\infty})_{rad}, \qquad t_2 = (t_{OB})_{part} + (t_{B\infty})_{rad}. \tag{9.8}$$

Thus

$$\Delta t \sim M/\gamma^2 \sim M \Delta\varphi (v^{-1}-1) \sim M \Delta\varphi/\gamma^2 \quad \Rightarrow \quad \Delta\varphi \sim 1 . \tag{9.9}$$

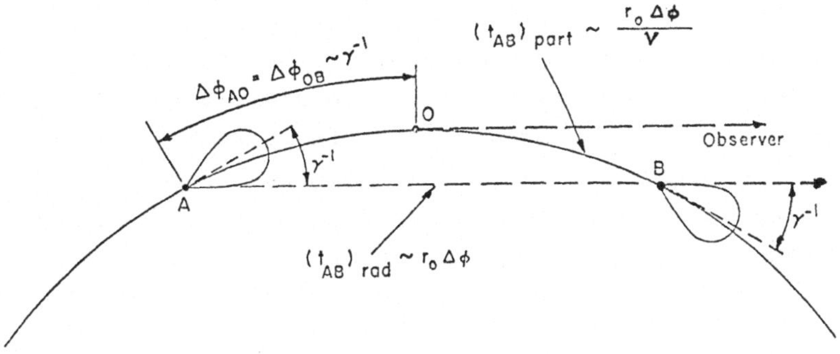

Figure 20(a). (from Chrzanowski & Misner 1974).
Illustration of Radiation emitted by an accelerated particle. Shown is a small arc of the particle's circular orbit and the emitted radiation, which is beamed into a narrow forward cone with half-width γ^{-1}. Choose the particular observer at infinity who is in the direction tangent to the orbit at point O. Owing to the beaming of radiation, this observer only sees radiation which is emitted between points A and B, where $\Delta\phi_{AO} = \Delta\phi \sim \gamma^{-1}$.

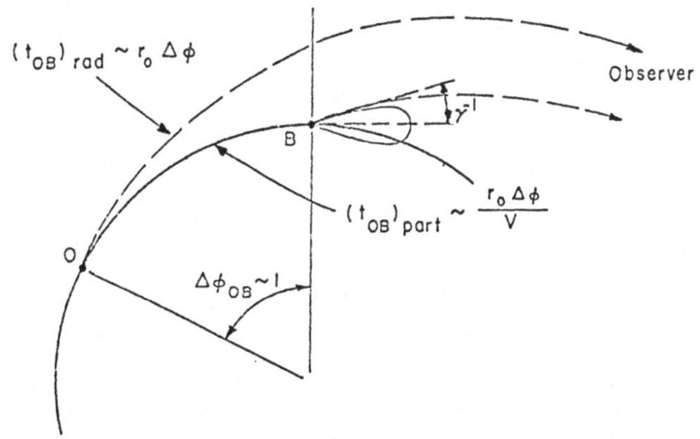

Figure 20(b). Radiation emitted by a geodesic particle. The radiation emitted at point O follows a null geodesic of the background geometry and reaches a particular observer at infinity. Photons emitted into the inner half of the forward cone before the particle is at point O cannot reach the observer, for they are captured by the black hole. Let B be the point where photons on the outer edge of the forward cone reach the given observer. Then, by construction, only radiation emitted between points O and B is seen by the observer. In Eq.(9.9) it is argued from $\omega_{crit} = \omega_0 \gamma^2$ that $\Delta\phi_{OB} \sim 1$.

There is no azimuthal beaming for radiation from particles moving relativistically on geodesics. A pulse will thus be seen by a distant observer even when the particle emits only a short burst of radiation into a forward cone which does not include the direction to the observer. It is only noted here as a not understood feature that for relativistic bremsstrahlung (small angle scattering) Peters (1970) and Matzner & Nutku (1973) get $E_{out} \sim \gamma^3$; therefore more energy is predicted from distant by-pass than from close revolutions of the black hole.

Amplitude factor

The amplitude factor given in Eq. (9.4) and Fig. 18 common for all three spectra surprisingly goes to zero as $1-(a/M)^2$ for black holes close to the extreme Kerr configuration ($a \rightarrow M$); i.e. there is a cutoff of radiation as seen from far away. Either the radiation does not leave the vicinity of the black hole when $a \rightarrow M$ (in which case it is absorbed by the black hole) or the particle ceases to radiate.

Here the latter possibility is supported by the following qualitative picture. A positive definite quantity which gives a measure of the strength of the gravitational field and hence of radiation is the "gravitational density" (Penrose 1960) $T_{abcd}u^a u^b u^c u^d/(u_e u^e)^2$. Here T_{abcd} is the Bel-Robinson tensor and u^a is the tangent to some observer's world line. In vacuum this quantity is positive unless the curvature vanishes or u^a becomes a (future-directed) null vector which points in one of the Principal Null Directions (PND) of the gravitational field. As Penrose (1960) remarks, "the gravitational PND are characterized by the fact that for observers travelling with a given velocity infinitesimally less than c, the gravitational density will be a minimum for those observers who travel approximately along a PND".

For $a \rightarrow M$ the null direction of the corotating circular photon orbit approaches a (repeated) PND of the Kerr geometry. Since this orbit represents approximately the particle orbit required for GSR, this is the situation mentioned by Penrose. Stewart & Walker (1973b) noticed that gravitational tidal forces which usually are $O(\gamma^2)$ are reduced by orders of magnitude to $O(1)$, for motion along a repeated PND. GSR is also of order γ^2 in general. One may therefore conclude: when the tangent vector of a worldlike of a particle becomes asymptotically close to a PND, kinematical effects like tidal forces & GSR will be minimized. This picture is consistent with the radiation cutoff for extreme Kerr.

9.2 Spin Dependence of Power Spectra

When the GSR spectra for the electromagnetic and gravitational case were obtained in 1972 (for the Schwarzschild geometry) the surprise was to find a dependence on frequency different from the scalar spectrum. The underlying expectation for employing scalar radiation had been that for high frequencies and high angular momentum modes the spin of the field would play only a negligible role for the spectrum. However, this turned out not to be the case. Chitre & Price (1972) first argued that the spin-dependence results from a break down of the geometrical optics approximation. But the validity of the WKB condition (8.43) guarantees the validity of geometrical optics in the radial direction. Indeed Chrzanowski & Misner (1974) applied high frequency methods to obtain all GSR spectra in the Kerr geometry.

Using geometrical optics arguments it can be seen that the spin-dependence arises from the different way of coupling of the source to the transverse field components. In the geometrical optics approximation the electromagnetic potential A_a and the gravitational metric perturbations h_{ab} (see MTW, Chap. 35) can be written as

$$A_a = e_a \Phi \ , \qquad h_{ab} = e_{ab} \Phi \ , \tag{9.10}$$

where Φ = (slowly varying function of position) $\times$ (rapidly varying phase) $= \mathfrak{A}\, e^{i\psi}$ is a solution of the scalar wave equation. The polarization vector e_a and the symmetric tracefree polarization tensor e_{ab} are orthonormal for the two transverse polarization states P, P'

$$e_a(P)e^a(P') = \delta_{PP'} \qquad e_{ab}(P)e^{ab}(P') = 2\,\delta_{PP'}, \tag{9.11}$$

are parallely propagated along $\psi_{,a} \equiv k_a$, and are transverse, $e_a k^a = e_{ab}k^b = 0$. Using

$$A_a^{out}(\ell m\omega P,x) = e_a(P)\,\phi^{out}(x,\ell m\omega)\,, \quad h_{ab}^{out}(\ell m\omega P,x) = e_{ab}(P)\,\phi^{out}(\ell m\omega,x) \tag{9.12}$$

in the general expressions for the spectra, Eqs. (8.70) and (8.86), shows that the quantities $e_a(P)u^a$ and $e_{ab}(P)u^a u^b$ have to be computed at the source of the radiation. Recalling (8.61) then yields

$$\frac{dP^{(1)}_{out}}{d\omega} = |e_a u^a|^2 \frac{dP^{(o)}_{out}}{d\omega} , \quad \frac{dP^{(2)}_{out}}{d\omega} = |e_{ab} u^a u^b|^2 \frac{dP^{(o)}_{out}}{d\omega} . \tag{8.13}$$

For the inner products in (9.13) obtains (Chrzanowski & Misner 1974)

$$|e_a u^a|^2 \sim \left(u^{\hat\varphi} \frac{k_\perp}{k_{\hat t}}\right)^2, \quad |e_{ab} u^a u^b|^2 \sim \left(u^{\hat\varphi} \frac{k_\perp}{k_{\hat t}}\right)^4 , \tag{9.14a}$$

where $u^{\hat\varphi}$ is the particle's 4-velocity component, $k_{\hat t}$ a component of the radiation momentum 4-vector, $k_\perp$ its momentum transverse to the particle motion, and all components are with respect to Bardeen's locally nonrotating frame. From the geodesic equations one has $u^{\hat\varphi} \sim \gamma \sim m^{1/2}_{crit}$ and $k_\perp / k_{\hat t} \sim \sin\varphi \sim \varphi$, where φ is the angle between the direction of the particle motion and a "typical" emitted photon. Such a photon is emitted with the half-width angle of the beam $\varphi \sim \Delta\vartheta = |m|^{-1/2}$. Hence (9.14) becomes

$$|e_a u^a|^2 \sim \left|\frac{m_{crit}}{m}\right| , \quad |e_{ab} u^a u^b|^2 \sim \left|\frac{m_{crit}}{m}\right|^2 , \tag{9.14b}$$

and

$$\frac{dP^{(s)}_{out}}{d\omega} = \left(u^{\hat\varphi} \frac{k_\perp}{k_{\hat t}}\right)^{2s} \frac{dP^{(o)}_{out}}{d\omega} = \left|\frac{\omega_{crit}}{\omega}\right|^s \frac{dP^{(o)}_{out}}{d\omega} \tag{9.15}$$

for $\omega \lesssim \omega_{crit}$. Similarly, the synchrotron spectra from accelerated particles in flat space (4.51) can be rewritten. There $m_{crit} \sim \gamma^3$ and $\Delta\vartheta \sim |m|^{-1/3}$ for $\omega < \omega_{crit}$, thus $|u| \sim |m_{crit}|^{1/3}$ and $k_\perp / |k| \sim |m|^{-1/3}$. Hence the spectra are related by

$$\frac{dP^{(s)}_{out}}{d\omega} = \left(|u| \frac{k_\perp}{|k|}\right)^{2s} \frac{dP^{(o)}_{out}}{d\omega} = \left|\frac{\omega}{\omega_{crit}}\right|^{-2s/3} \frac{dP^{(o)}_{out}}{d\omega} , \tag{9.16}$$

in accordance with Eq. (4.51).

9.3 Limitations of Test Particle GSR

Several restrictions limit the physical applicability of results derived within the test particle approximation and under the assumption of geodesic motion. In the framework of gravitational perturbation theory one has the possibilities of assuming both point particles and geodesic motion, dropping one of the assumptions or renouncing both. Finally, the exact treatment assumes the full theory without any such conditions. Some of these cases have been employed, GSR is an example for the first (the simplest) case. There the validity of the computations is strongly limited.

Radiation Damping

Geodesic motion and emission or radiation are strictly speaking incompatible since geodesic motion implies that the total energy radiated from a circular orbit is infinite. It provides a reasonable picture only as long as the perturbation of the geodesic orbits by radiation damping is very small. However, since the general two-body problem is as yet unsolved in GR, one has to rely on expansion schemes within the limits of slow or fast motion, weak fields and small mass ratios. Even with these approximations, radiation damping causes considerable difficulties. This question is being investigated within the framework of approximation schemes by Burke, Havas, Rosenblum, and Rudolph. Newman and coworkers have given a not well understood but exact treatment on $\mathcal{J}^+$, while Aichlburg & Beig have given a fully understandable treatment but only in the context of a model theory.

Conservation Laws

Even when particles move along damped orbits in the field of "large" mass, one commits a violation of momentum conservation when the large mass is kept fixed. For "small" particles this effect may even be of the same order as radiation damping. This in fact should cause "spurious" dipole radiation which, however, does not show up in the approximations considered here.

Extended Bodies

Physically, the motion of extended bodies around black holes has to be considered since a point particle on a geodesic should not radiate and since the "point particle" concept seems to be incompatible with GR. Fishbone (1973, 1975) has kept the assumption of (circular) geodesic motion but considered ellipsoidal, homogeneous, incompressible small fluid bodies orbiting a Kerr black hole. He studied the relativistic Roche problem, i.e. the equilibrium condition for the existence of these bodies in stable and GSR orbits, before tidal forces disrupt them.

However, rather little is known about how extended bodies move in GR (for a special case see Geroch & Jang 1975). A discussion of the motion of extended bodies in GR has been given by Dixon, but an application of this formalism to the questions of this section is still to be awaited.

9.4 Some Open Questions in Gravitational Radiation Theory

During the course of this article some problems have arisen which seem to constitute unsolved problems or have not been treated satisfactorily, although several of them are currently being worked on.

1. Reformulate the Regge-Wheeler perturbation theory in a gauge-invariant manner in the style of Moncrief (1974abc).

2. Do there always exist metric perturbations for any given curvature perturbation?

3. Starting from GHP formalism, give an intrinsic characterization of the splitting of a corresponding metric into unperturbed and perturbed part.

4. Formulate the initial value problem for spacetime perturbations. (see O'Murchada & York 1974).

5. Decouple & separate the perturbation equations in the Kerr-Newman geometry (for special cases see Chitre 1975; Gerlach 1974; Bose & Wang 1974, Bose 1975; for Reissner-Nordstrøm: Zerilli (1974) and Moncrief (1974ab); Gravitational perturbations are trivial: Stewart & Walker (1973)).

6. Study scattering problem with complex potential as it is the case in Teukolsky's radial master equation. (For axisymmetric pertubations: Chandrasekhar & Detweiler, 1975.)

7. Show strong completeness of eigenfunctions of Teukolsky's equations for complex frequencies (thereby more or less completing the sta-

bility proof of the Kerr geometry; see Stewart 1975).

8. Show the existence of Chrzanowski & Misner's (1974) asymptotically factorized Green's function solutions (a) asymptotically on $\mathcal{J}^+$, (b) everywhere in the space of outer communications.

9. Find the relation of solutions of linearized (perturbation) equations to solutions of the exact equations ("linearization stability"). (See Fisher & Marsden 1974, Choquet-Bruhat 1969, Brill & Deser 1973, Moncrief 1974d).

10. Give an exact solution for a spacetime with both source and radiation, or at least one which goes beyond the linear approximation.

11. Develop a formalism for gravitational perturbation theory within the frame-work of Regge-Calculus (section 5.7).

12. Show that asymptotically flat spacetimes are linearization stable in the sense of Fisher & Marsden.

13. Give a method to solve four- and five-term recursion relations (Sec. 6.12) generalizing the method of infinite fractions as it is known for three-term recursion relations.

ACKNOWLEDGMENTS

I am grateful to my colleagues in the Munich group for their help during this work. Especially I thank John Stewart for many enlightening discussions and advice. Numerous improvements were caused by Jürgen Ehlers's critical reading of the manuscript. Garry Ludwig proofread the English and Bernd Schmidt provided creative discouragement. I thank Paul Chrzanowski who was very helpful with calculations in Chap. VIII, Victor Hamity, Peter Kafka, Michael Streubel and Martin Walker for several useful discussions, and Rosita Strasser for typing the final manuscript.

APPENDIX A

We list the notations used in the present work. As for abbreviations we occasionally use GSR for Geodesic Synchrotron Radiation, GR for the Theory of General Theory, LNRF for Bardeen's locally nonrotating family of observers in the Kerr geometry and SP for Stokes Parameters. Also, MTW refers to the book by Misner, Thorne & Wheeler (1973).

Indices from the first half of the latin alphabet (a,b,...g,h) assume the values (0,1,2,3), the second half of the alphabet (i,j,k, ...y,z) run through (1,2,3). The time component of a vector is the component with the index 0. Throughout [...] and (...) denote the totally skew and symmetric part, respectively; e.g. $A_{[ab]} = \frac{1}{2}(A_{ab} - A_{ba})$, $A_{(ab)} = \frac{1}{2}(A_{ab} + A_{ba})$. $\Psi_{a;b} \equiv \nabla_b \Psi_a$ denotes the covariant derivative and $\Psi_{,a} = \partial\Psi/\partial x^a$, $\partial_a = \partial/\partial x^a$, $\partial_a\partial^a = \Box$. Three-dimensional vectors are underligned.

The signature of spacetime is (- + + +). The components of the Minkowski metric (in Cartesian coordinates) are η = diag (-1,+1,+1,+1). Differentiability is assumed throughout to the order required. The physical units and dimensions are chosen so that $G = c = 1$.

Penrose's notiation for the conformal boundary of asymptotically flat spacetimes and its parts are used; i.e. future null infinity $\mathcal{J}^+$, past null infinity $\mathcal{J}^-$, spatial infinity I^0, future timelike infinity I^+, past timelike infinity I^-.

APPENDIX B

High Multipole Spin-Weighted Spherical Harmonics

In Eq. (8.56) only the leading term of even parity ($k = \ell - m$ = even) was given since the relation

$$|{}_sZ^{\omega}_{\ell m} \;(\pi/2,0)|^2 = (1-a^2\omega_0^2)^{1/4}\,|{}_sY_{\ell m}(\pi/2,0)|^2 \tag{B.1}$$

is valid only up to $O(m^0)$. However, if one wants to know more precisely the contribution of the odd parity terms to the power in Eq. (8.81) and (8.91) one has to include also the leading terms for odd k. For this purpose the high-frequency approximation of the ${}_sY_{\ell m}(\theta,\varphi)$ is required. To find the leading terms in both parities take ℓ=m for even, ℓ=m+1 for odd parity. In this case the ingredients required to obtain the ${}_sY_{\ell m}$ using their defining relations (6.60) are

$$Y_m{}^m(\theta,\varphi) = (-1)^m(4\pi^3)^{-1/4}\, m^{1/4}\, \sin^{|m|}\theta\; e^{im\varphi}\,(1+1/4m),$$

$$\partial_\theta Y_m{}^m(\theta,\varphi) = (-1)^m(4\pi^3)^{-1/4}\, m^{5/4}\sin^{|m|-1}\theta\,\cos\theta\; e^{im\varphi}\,(1+1/4m), \tag{B.2}$$

$$\partial_\varphi Y_m{}^m(\theta,\varphi) = i(-1)^m(4\pi^3)^{-1/4}\, m^{5/4}\sin^{|m|}\theta\; e^{im\varphi}\,(1+1/4m);$$

$$Y_{m+1}{}^m(\theta,\varphi) = (-1)^m\,\pi^{-3/4}\, m^{3/4}\,\sin^{|m|}\theta\,\cos\theta\; e^{im\varphi}\,(1+11/8m),$$

$$\partial_\theta Y_{m+1}{}^m(\theta,\varphi) = (-1)^m\,\pi^{-3/4}\, m^{7/4}\,\sin^{|m|-1}\theta\,(\cos^2\theta-\frac{\sin^2\theta}{m})\, e^{im\varphi}(1+11/8m), \tag{B.3}$$

$$\partial_\varphi Y_{m+1}^m(\theta,\varphi) = i(-1)^m\,\pi^{-3/4} m^{7/4}\,\sin^{|m|}\theta\,\cos\theta\; e^{im\varphi}\,(1+11/8m).$$

Inserting these expressions appropriately into Eq. (6.60) one obtains for $\ell = m$

$$_{\pm 1}Y_{m,m}(\theta,\varphi) = (-1)^{m+1}(4\pi^3)^{-1/4}\, m^{1/4}\,\sin^{|m|-1}\theta\,(\cos\theta\mp 1)\; e^{im\varphi},$$

$$_{\pm 2}Y_{m,m}(\theta,\varphi) = (-1)^m(4\pi^3)^{-1/4}\, m^{1/4}\,\sin^{|m|}\theta\Big[1+\frac{2\cot\theta}{\sin\theta}(\cos\theta\mp 1)\Big]\, e^{im\varphi}, \tag{B.4}$$

and for $\ell = m+1$

$$_{\pm 1}Y_{m+1,m}(\theta,\varphi) = (-1)^m\pi^{-3/4}\, m^{3/4}\sin^{|m|}\theta\Big[\cos\theta(\cot\theta\mp 1) - \frac{1}{m}(\sin\theta - \frac{11}{8}\cos\theta(\cot\theta\mp 1))\Big]e^{im\varphi}, \tag{B.5}$$

$$_{\pm 2}Y_{m+1,m}(\theta,\varphi) = (-1)^m\pi^{-3/4}\, m^{3/4}\sin^{|m|}\theta\Big[2\cot^2\theta(\cos\theta\mp 1) + \cos\theta \pm 2/m + \frac{8}{m}\cot^2\theta(6\cos\theta\pm 5) + \frac{11}{8m}\cos\theta\Big]e^{im\varphi}.$$

In the equatorial plane ($\theta = \pi/2$) the odd parity multipole modes in (B.5) are of order $m^{-1/4}$ hence by one order of magnitude smaller in m than the even parity modes in (B.4) which was claimed in Eq. (8.56).

REFERENCES

Alfaro, V.de, and Regge, T. 1965. Potential Scattering (Amsterdam: North-Holland).

Anile, S.M. 1973, Ph.D. Thesis, Oxford University (unpublished).

Anile, S.M., and Breuer, R.A. 1974, Ap. J., **189**, 39-49, "Gravitational Stokes Parameters".

Bardeen, J.M. 1970a, Nature, **226**, 64-65, "Kerr metric black holes".

Bardeen, J.M. 1970b, Ap. J., **161**, 103-109, "Stability of circular orbits in stationary, axisymmetric spacetimes".

Bardeen, J.M. 1970c, Ap. J., **162**, 71-95, "A variational principle for rotating stars in general relativity".

Bardeen, J.M., Press, W.H., and Teukolsky, S.A. 1972, Ap. J., **178**, 347-369, "Rotating black holes: Locally nonrotating frames, energy extractions and scalar synchrotron ratiation".

Bardeen, J.M., and Press, W.H. 1973, J. Math. Phys., **14**, 7-19, "Radiation fields in the Schwarzschild background".

Bertotti, B., and Cavalieri, A. 1971, Nuov. Cim., **2B**, 223-235, "Gravitational waves and cosmology".

Billing, B., Kafka, P., Maischberger, K., Meyer, F., and Winkler, W. 1975, Nuov. Cim. Lett., **12**, 111-116, "Results of the Munich-Frascati gravitational-wave experiment".

Bondi, H., and Gold, T. 1955, Proc. Roy. Soc. A, **229**, 416-424, "The field of a uniformly accelerated charge, with a special reference to the problem of gravitational radiation."

Bose, S.K., and Wang, M.Y. 1974, Phys. Rev. D, **10**, 1675-1677, "Stationary, axisymmetric perturbations of charged black holes."

Bose, S.K. 1975, "Studies in the Kerr-Newman metric" (preprint).

Bouwkamp, C.J. 1950, Phillips. Res. Rep., **5**, 401-422, "On the characteristic values of spheroidal wave functions".

Bourassa, R.R., and Kantowski, R. 1975, Ap. J., **195**, 13-21, "The theory of transparent gravitational lenses".

Boyer, R.H., and Lindquist, R.W. 1967, J. Math. Phys., **8**, 265-281, "Maximal analytic extension of the Kerr metric."

Breuer, R.A., Tiomno, J., and Vishveshwara, C.V. 1972, Nuov. Cim. Lett., 4, 857-860; 1975, Nuov. Cim., 25B, 851-870, "Polarization of gravitational synchrotron radiation".

Breuer, R.A., Chrzanowski, P.L., Hughes, H.G. III, and Misner, C.W. 1973, Phys. Rev. D, 8, 4309-4319, "Geodesic synchrotron radiation".

Breuer, R.A., Ruffini, R., Tiomno, J., and Vishveshwara, C.V. 1973, Phys. Rev. D, 7, 1002-1007, "Vector and tensor radiation from Schwarzschild circular geodesics".

Breuer, R.A., and Vishveshwara, C.V. 1973, Phys. Rev. D, 7, 1008-1017, "Polarization of synchrotron radiation from relativistic Schwarzschild geodesics".

Breuer, R.A., and Ryan, M.P. 1975, "High- and low frequency expansions of spin-weighted spheroidal harmonics", to be published.

Brill D.R. 1973, in Relativity, Astrophysics and Cosmology, W. Israel (ed.) (Dordrecht-Holland: D.Reidel Publ. Co.), p. 127-152.

Brill, D.R., Chrzanowski, P.L., Pereira, C.M., Fackerell, E.D., and Ipser, J.R. 1972, Phys. Rev. D, 5, 1913-1915, "Solution of the scalar wave equation in a Kerr background by separation of variables".

Brill, D.R., and Deser, S. 1973, Comm. Math. Phys., 32, 291-304, "Instability of closed spaces in general relativity".

Burke, W.L. 1970, Phys. Rev. A, 2, 1501-1505, "Runaway solutions: remarks on the asymptotic theory of radiation damping".

Burke, W.L. 1971, J. Math. Phys., 12, 401-418, "Gravitational radiation damping of slowly moving systems calculated using matched asymptotic expansions."

Campbell, G.A., and Matzner, R.A. 1973, J. Math. Phys., 14, 1-6, "A model for peaking of galactic gravitational radiation".

Carter, B. 1968, Phys. Rev., 174, 1559-1571 "Global structure of the Kerr family of gravitational fields".

Carter, B. 1973, Comm. Math. Phys., 30, 261-286 "Elastic perturbation theory in general relativity and a variation principle for a rotating solid star".

Carter, B., and Quintana, H. 1972, Proc. Roy. Soc. A, 331, 57-83, "Foundations of general relativistic high-pressure elasticity theory".

Chandrasekhar, S. 1950, Radiative Transfer (Oxford: Clarendon Press).

Chandrasekhar, S. 1975, Proc. Roy. Soc. A (in press) "On the equation governing the perturbations of the Schwarzschild black holes".

Chandrasekhar, S., and Detweiler, S. 1975, Proc. Roy. Soc. A (in press) "On the equations governing the axisymmetric perturbations of the Kerr black hole".

Choquet-Bruhat, Y. 1969, Comm. Math. Phys., 12, 16-35, "Construction de solutions radiatives approchées de équations d'Einstein".

Choquet-Bruhat, Y., and Deser, S. 1973, Ann. Phys., 81, 165-178, "On the stability of flat space".

Chitre, D.M., and Price, R.H., Phys. Rev. Lett., 29, 185-188, "Nature of synchrotron radiation."

Chitre, D.M., Price, R.H., and Sandberg, V.D. 1973, Phys. Rev. Lett., 31, 1018-1022 "Electromagnetic radiation from an unmoving charge"; Phys. Rev. D., 11, 747-759 "Electromagnetic radiation due to space-time oscillations".

Chitre, D.M. 1975, Phys. Rev. Lett. 11, 760-762, "Electromagnetic radiation by gravitational perturbations of a charged rotating black hole".

Chrzanowski, P.L. 1973, Gravitational Synchrotron Radiation, Ph.D. thesis, Univ. of Maryland (unpublished).

Chrzanowski, P.L., and Misner, C.W. 1974, Phys. Rev. D, 10, 170-1721 "Geodesic synchrotron radiation in the Kerr geometry by the method of asymptotically factorized Green's functions".

Chrzanowski, P.L. 1975, Phys. Rev. D., 11, 2042-2062, "Vector and metric perturbations of a rotating black hole".

Clarke, D., and Graininger, J.F. 1971, Polarized Light and Optical Measurement, Int. Ser. of Monographs in Nat. Phil. Vol 35 (Pergamon Press).

Cocke, W.J., and Holm, C.A. 1972, Nature, 240, 161-162, "Lorentz transformation properties of Stokes parameters."

Cohen, J.M., and Kegeles, L.S. 1974, Phys. Rev. D, 10, 1070-1084 "Electromagnetic fields in curved spaces: a constructive approach".

Collins, P.A., and Williams, R.M. 1972, Phys. Rev. D, 5, 1908-1912, "Application of Regge-calculus to the axially symmetric initial-value problem in general relativity".
1973, Phys. Rev. D, 7, 965-971, "Dynamics of the Friedman universe using Regge calculus"; 1975, Univ. of Birmingham preprint "A Regge calculus model for the Tolman universe".

Davis, M., Ruffini, R., Press, W.H., and Price, R.H. 1971, Phys. Rev. Lett., 27, 1466-1469 "Gravitational radiation from a particle falling radially into a Schwarzschild black hole".

Davis, M., Ruffini, R., Riomno, J., and Zerilli, F. 1972, Phys.Rev.Lett. 28, 1352-135 "Can gravitational synchrotron radiation exist?".

De Felice, F. 1968, Nuov. Cim., 57B, 351-388, "Equatorial geodesic motion in the gravitational field of a rotating source".

Desai, C.W., and Abel, J.F. 1972, Introduction to the Finite Element Method (New York: Van Nostrand Reinhold Co.).

Detweiler, S. 1975, "Electromagnetic perturbations of a rotating black hole", to be published.

Doroshkevich, A.G., Novikov, I.D., and Polnarev, A.G. 1973, Sov. Phys. JETP, **36**, 816-820, "Gravitational synchrotron radiation".

Eardley, D.M., Lee, D.L., Lightman, A.P., Wagoner, R.V. and Will, C.M. 1973, Phys.Rev.Lett., **30**, 884-886; 1973, Phys.Rev., **D8**, 3308-3321, "Gravitational wave observations as a tool for testing relativistic gravity".

Edelstein, L.A., and Vishveshwara, C.V. 1970, Phys. Rev. D, **1**, 1314-3517, Differential equations for perturbations on the Schwarzschild metric".

Ehlers, J. 1973, in Relativity, Astrophysics and Cosmology, W. Israel (ed.) (Dordrecht-Holland: D. Reidel Publ. Co.) p. 1-125.

Einstein, A. 1913, Phys. Z., **14**, 1249-1266, "Zum gegenwärtigen Stand des Gravitationsproblems".

Einstein, A. 1936, Science, **84**. 506.

Ellis, G.F.R. 1971, in General Relativity and Cosmology (New York & London: Ac. Press) p. 104.

Erdelýi, A., Magnus, W., Oberhettinger, F., and Triomi, F.G. 1955, Higher Transcendental Functions (New York: McGraw-Hill Co.) Vol. III.

Erdelýi, A. 1956, Asymptotic Expansions (New York: Dover).

Fischer, A.E., and Marsden, J.E. 1973, Bull. Am. Math. Soc., **79**, 997-1003, "Linearization stability of the Einstein equations".

Fishbone, L.H. 1973, Ap. J., **185**, 43-67, "The relativistic Roche problem. I. Equilibrium theory for a body in equatorial, circular orbit around a Kerr black hole".

Fishbone, L.H. 1975, Ap. J., **195**, 499-505, "The relativistic Roche problem. II. Stability theory.

Flammer, C. 1957, Spheroidal Wave Functions (Standford: Univ. Press).

Friedman, B. 1965, Principles and Techniques of Applied Mathematics (New York, London: Wiley).

Fulton, T., and Rohrlich, F. 1960, Ann. Phys. (N.Y.), **9**, 499-517, "Classical radiation from a uniformly accelerated charge".

Gel'fand, I.M., Minlos, R.A., and Shapiro, Z. Ya. 1963, Representation of the Rotation and Lorentz Groups and their Applications, (New York: Macmillan Co.).

Gerlach, U.H. 1974, Phys.Rev.Lett., **32**, 1023-1025,"Beat frequency oscillations near charged black holes;1974, "Faraday rotation near charged black holes" (preprint).

Geroch, R. 1969, Comm. Math. Phys., **13**, 180-193, "Limits of spacetimes".

Geroch, R., and Pong Soo Jang 1975, J. Math. Phys., **16**, 65-67, "Motion of a body in general relativity".

Geroch, R., Held, A., and Penrose, R. 1973, J. Math. Phys., **14**, 874-881, "A spacetime calculus based on pairs of null directions".

Green, R.D., Schücking, E.L., and Vishveshwara, C.V. 1975, J. Math. Phys., **16**, 153-157, "The rest frame in stationary spacetimes with axial symmetry".

Gum, C.S., Kerr, F.J., and Westerhout, G. 1960, M.N.R.A.S., **121**, 132-149, "A 21-cm determination of the principal plane of the galaxy".

Gunn, J.E., and Press, W.H. 1973, CalTech OAP-320, "Method for detecting a cosmological density of condensed objects."

Hartle, J.B., Wilkins, D.C. 1974, Comm. Math. Phys., **38**, 47-64, "Analytic properties of the Teukolsky equation".

Hawking, S.W., and Ellis, G.R.F. 1973, The Large Scale Structure of Space-Time (Cambridge: Univ. Press).

Hawking, S.W. 1974, Nature, **248**, 30-31, "Black hole explosions?"

Herlt, E., and Stephani, H. 1975, Univ. of Jena preprint, "Wave optics of the spherical lens".

Hughes, H.G. III, 1974, Ann. Phys. (N.Y.), **80**, 463-478, "The gravitational searchlight effect from synchrotron radiation in the Schwarzschild geometry".

Isaacson, R.A. 1968, Phys. Rev., **166**, 1263-1271, "Gravitational radiation in the limit of high frequency, I: The linear approximation and geometrical optics"; ibid., 1272-1280 "II: Nonlinear terms and the effective stress tensor".

Jackson, J.C. 1973, Nature, **241**, 513-515, "Gravitational waves and relativistic disks".

Jackson, J.D. 1967, Classical Electrodynamics (New York: Wiley & Sons).

Janis, A.I., and Newman, E.T. 1965, J. Math. Phys., **6**, 902-817, "Structure of gravitational sources."

Jauch, J.M., and Rohrlich, F. 1955, The Theory of Photons and Electrons (Cambridge, Mass.: Addison-Wesley).

Jones, R.C. 1956, J. Opt. Soc. Am., **46**, 126-131, "New calculus for the treatment of optical systems. VIII. Electromagnetic theory".

Kafka, P. 1970, Nature, **226**, 436-439, "Gravars".

Kafka, P., and Meyer, F. 1972, Mitt. Astr. Ges., **31**, 129-132, "Stand und Aussichten der Gravoastronomie".

Kerr, R.P. 1963, Phys. Rev. Lett., **11**, 237-238, "Gravitational field of a spinning mass as an example of algebraically special metrics".

Kinnersley, W. 1969, J. Math. Phys., **10**, 1195-1203, "Type D vacuum metrics".

Kraus, J.D. 1966, Radio Astronomy (New York: McCraw-Hill).

Landau, L.D., and Lifshitz, E.M. 1962, Classical Theory of Fields (Reading, Mass.: Addison-Wesley).

Lawrence, J.K. 1971, Phys. Rev. D, **3**, 3239-3240, "Focussing of gravitational radiation by the galactic core".

Lawrence, J.K. 1972, Ap. J., **171**, 483-484, "Intensification of gravitational radiation by a massive rotator".

Lawrence, J.K. 1971, Phys. Rev. D, **7**, 2275-2277, "Focussing of gravitational radiation in the galactic plane".

Liebes, S. 1964, Phys. Rev. B, **133**, 835-844, "Gravitational lenses".

Lindquist, R.W. 1966, Ann. Phys. (N.Y.), **37**, 487-518, "Relativistic transport theory".

Ludwig, G. 1969, Am. J. Phys., **37**, 1225-1238, "Classification of electromagnetic and gravitational fields".

Ludwig, G. 1974, J. Math. Phys., **15**, 928-932, "On the geometrization of neutrino fields I".

Lynden-Bell, D. 1969, Nature, **223**, 690-694, "Galactic nuclei as collapsed old quasars".

Lynden-Bell, D., and Rees, M.J. 1971, M.N.R.A.S., **152**, 461-475, "On quasars, dust, and the galactic center".

McCallum, M.A.H., and Taub, A.H. 1973, Comm. Math. Phys.,**30**, 153-169, "The averaged Lagrangian and high-frequency gravitational waves".

Madore, L. 1972, Comm. Math. Phys., **27**, 291-302, "The dispersion of gravitational waves".

Madore, L. 1973, Comm. Math. Phys., **30**, 335-340, "The absorption of gravitational radiation by a dissipative fluid".

Magnus, W., Oberhettinger, F., and Soni, R.P. 1966, Formulas and Theorems for the Special Functions of Mathematical Physics Berlin, Heidelberg, New York: Springer Verlag).

Matzner, R. 1968, J. Math. Phys., **9**, 163-170, "Scattering of massless scalar waves by a Schwarzschild singularity".

Matzner, R.A., and Nutku, Y. 1974, Proc. Roy. Soc., **336A**, 285-305 "On the method of virtual quanta and gravitational radiation".

Meixner, J. 1948, Z. Math. Mech., **28**, 304-310, "Asymptotische Entwicklungen der Eigenwerte und Eigenfunktionen der Differentialgleichungen der Sphäroid-Funktionen und der Mathienschen Funktionen".

Meixner, J., and Schäfke, F.W. 1954, Mathieusche Funktionen und Spheroidfunktionen (Berlin: Springer Verlag).

Misner, C.W. 1969, in Astrophysics and General Relativity, M. Chretien, S. Deser & J. Goldstein (eds.), Vol I (New York: Gordon & Breach), "Gravitational Collapse".

Misner, C.W. 1972a, Phys. Rev. Lett., **28**, 994-997, "Interpretation of gravitational wave observations".

Misner, C.W. 1972b, Bull. Am. Phys. Soc., **17**, 472, "Stability of Kerr black holes against scalar perturbations".

Misner, C.W., Breuer, R.A., Brill, D.R., Chrzanowski, P.L., Hughes, H.G.III, and Pereira, C.M. 1972, Phys. Rev. Lett., **28**, 998-1001, "Gravitational synchrotron radiation in the Schwarzschild geometry".

Misner, C.W., Thorne, K.S., and Wheeler, J.A. 1973, Gravitation (San Francisco: W.H. Freeman & Co.).

Misner, C.W. 1974, IAU Symposium Nr. 64, C. de Witt (ed.), p. 3 (Dordrecht: Reidel Publ. Co.), "Mechanisms for the emission and absorption of gravitational radiation".

Moncrief, V. 1974a, Phys. Rev. D, **9**, 2707-2709, "Odd-parity perturbations of a Reissner-Nordstrøm black hole".

Moncrief, V. 1974b, Phys. Rev. D, **10**, 1057-1053, "Stability of a Reissner-Nordstrøm black hole."

Moncrief, V. 1974c, Ann. Phys. (N.Y.), **88**, 323-342, "Gravitational perturbations of spherically symmetric systems I. The exterior problem".

Moncrief, V. 1975, J. Math. Phys., **16**, 493-498, "Spacetime symmetries and linearization stability of the Einstein equations I".

Morse, P.M., and Feshbach, H. 1963, Methods in Theoretical Physics (New York: McGraw-Hill).

Müller zum Hagen, H., Yodzis, P., and Seifert, H.-J. 1973, Comm. Math. Phys., **34**, 135-148; **37**, 19-40, "On the occurrence of naked singularities in general relativity. I,II".

Newman, E., and Penrose, R. 1962, J. Math. Phys., **3**, 566-578, "An approach to gravitational radiation by a method of spin-coefficients".

Newman, E., and Unit, T. 1962, J. Math. Phys., **3**, 891-901, "Behaviour of asymptotically flat empty spaces".

O'Murchada, N., and York, J. 1974, Phys. Rev. D, **10**, 437-446, "Initial-value problem of general relativity. II. Stability of solutions of the initial-value equations".

O'Neill, E.L. 1962, Introduction to statistical optics (Reading, Mass.: Addison-Wesley).

Papapetrou, A. 1954, Math. Nachr., **12**, 129 et 143; Z.Phys., **139**, 518.

Penrose, R. 1969, Rev. Nuov. Cim., **1** (Numero Speciale), 252-276, "Gravitational collapse: the role of general relativity".

Penrose, R. 1960, Ann. Phys. (N.Y.), **10**, 171-201, "A spinor approach to general relativity".

Penrose, R. 1964, in Relativity, Groups & Topology, B. De Witt & De Witt (eds.) (New York: Gordon & Breach) "Conformal treatment of infinity".

Penrose, R. 1972, Sci. Am., **226**, 38, "Black Holes"; see also Nature, 236, 277-380.

Peters, P.C. 1970, Phys. Rev. D, **1**, 1559-1571, "Relativistic gravitational bremsstrahlung".

Pirani, F.A.E. 1959, Proc.Roy.Soc., A252, 96-101,"The gravitational field of a fast moving particle."

Pirani, F.A.E. 1965, "Introduction to gravitational radiation theory", in Lectures on GR, Brandeis (Englewood Cliffs, N.J.: Prentice Hall).

Poincaré, H. 1892, Theorie Mathematique de la Lumière II, Chap. 12 (Paris).

Press, W.H. 1972, Ap. J., **175**, 243-252, "Time evolution of a rotating black hole immersed in a static field".

Press, W.H., and Thorne, K.S. 1972, Ann. Rev. Astron. Astrophys., 10, 335-374 "Gravitational wave astronomy".

Press, W.H., and Teukolsky, S.A. 1972, Nature, **238**, 211-212, "Floating orbits, superradiant scattering and the black hole bomb".

Press, W.H., and Teukolsky, S.A. 1973, Ap. J., **185**, 649-673 "Perturbations of a rotating black hole II. Dynamical stability of the Kerr metric"; Ap. J., **193**, 443-461 "Perturbations of a rotating black hole III. Interaction of the hole with gravitational and electromagnetic radiation".

Price, R.H. 1972a, Phys. Rev. D, **5**, 2419-2438, "Nonspherical perturbations of relativistic gravitational collapse, I: Scalar and gravitational perturbations".

Price. R.H. 1972b, Phys. Rev. D, **5**, 2439-2454, "Nonspherical perturbations of relativistic gravitational collapse, II. Integer-spin, zero-rest-mass fields".

Price, R.H., and Sandberg, V.D. 1973, Phys. Rev. D, **8**, 1640-1644, "Role of constraining forces for ultrarelativistic particle motion as a source of gravitational radiation".

Regge, T., and Wheeler, J.A. 1957, Phys. Rev., **108**, 1063-1069, "Stability of a Schwarzschild singularity".

Regge, T. 1961, Nuov. Cim., **19**, 558-571, "General relativity without coordinates".

Robinson, I. and Robinson, J.R. 1972, in General Relativity, papers in honour of J.L. Synge (Oxford: Clarendon Press).

Rohrlich, F. 1963, Ann. Phys. (N.Y.), 22, 169-191, "The principle of equivalence".

Ruffini, R. and Wheeler, J.A. 1971, The Significance of Space Research for Fundamental Physics, A.F. Moore & V. Hardy (eds.) (Paris: ESRO)

Ruffini, R., Tiomno, J., and Vishveshwara, C.V. 1972, Nuov. Cim. Lett. 3, 211-215, "Electromagnetic fields of a particle moving in a spherically symmetric black hole background".

Ryan, M.P., 1972, Ap. J., 177, L79-L84, "Is the existence of a galaxy evidence for a black hole at its center?"

Ryan, M.P. 1974, Phys. Rev. D, 10, 1736-1740, "Teukolsky equation and Penrose wave equation".

Sachs, R. 1964, in General Relativity, Groups & Topology, B. De Witt & C. De Witt (eds.) (New York: Gordon & Breach).

Schiff, L. 1968, Quantum Mechanics (New York: McGraw-Hill).

Schild, A. 1948, Phys. Rev., 4, 414-415, "Discrete Spacetime and integral Lorentz transformations".

Silberstein, L. 1936, Discrete Space-Time, Univ. of Toronto Studies, Physics Series.

Sciama, D.W. 1969, Nature, 224, 1263-1267, "Is the galaxy losing mass on a time scale of a billion years?"

Sciama, D.W., Field, G.F., and Rees, M.J. 1969, Phys. Rev. Lett, 23, 1514-1515, "Upper limit to radiation of mass energy derived from expansion of galaxy".
See also Field, G.F., Rees, M.J., and Sciama, D.W. 1969, Comm. Astr. Space Sci., 1, 187.

Shurcliff, W.A., and Ballard, S.S. 1964, Polarized Light (Princeton: Van Nostrand).

Slusher, R.E., and Tyson, J.A. 1973, Nature, 243, 25, "Search for infrared anomalies associated with gravitational events at the galactic centre".

Sokolov, A.A., and Ternov, I.M. 1968, Synchrotron Radiation (Oxford: Pergamon Press)

Sorkin, R. 1975, CalTech OAP-374, "The time evolution problem in Regge calculus", CalTech OAP-393, "The electromagnetic field on a simplectic net".

Starobinsky, A.A. 1973, Zh. Exp. i Teor. Fiz., 64, 48 (transl. in Soviet. Phys. JETP, 37, 28) "Amplification of waves during reflection from a rotating black hole".

Starobinsky, A.A., and Churilov, S.M. 1973, Zh. Exp. i Teor. Fiz., 65, 3, "Amplification of electromagnetic and gravitational waves scattered by a rotating black hole".

Stewart, J.M. 1972, preprint; see also Stewart & Walker (1973a).

Stewart, J.M. 1973, Phys. Lett., A44, 499-500, "Stationary perturbations of the Kerr solution".

Stewart, J.M. 1975, Proc. Roy. Soc. A, 334, 51-64, "On the stability of Kerr's spacetime".

Stewart, J.M. 1975, Proc. Roy. Soc. A, 334, 65-76, "Global solutions of ordinary differential equations with polynomial coefficients".

Stewart, J.M., and Walker, M. 1973a, Springer Tracts, 69, (Berlin: Springer Verlag), "Black holes: the outside story".

Stewart, J.M., and Walker, M. 1973b, Comm. Math. Phys., 29, 43-47, "Tidal accelerations in general relativity".

Stewart, J.M. and Walker, M. 1974, Proc.Roy.Soc.,A341,49-74,"Perturbations of spacetimes in general relativity".

Stokes, G.G. 1852, Trans. Camb. Phil. Soc., 9, 399-418, "On the composition and resolution of streams of polarized light from different sources".

Teukolsky, S. 1972, Phys. Rev. Lett., 29, 1114-1118, "Perturbations of a rotating black hole: Separable wave equations for gravitational and electromagnetic perturbations".

Teukolsky, S. 1973, Ap. J., 185, 635-647, "Perturbations of a rotating black hole, I: Fundamental equations for gravitational, electromagnetic, and neutrino-field perturbations".

Thorne, K.S. 1969, Ap. J., 158, 1-16, "Nonradial pulsations of general relativistic stellar models, IV: The weak-field limit".

Trautmann, A. 1962, "Conservation laws in general relativity", Chap. V in Gravitation: an introduction into current research (ed.) (London, New York: Wiley & Sons).

Trautmann, A. 1965, "Foundations and current problems of general gelativity theory" in Lectures on General Relativity, Brandeis (Englewood Cliffs, N.J.: Prentice-Hall).

Tyson, J.A., and Douglass, D.H. 1972, Phys. Rev. Lett., 28, 991-994, "Response of a gravitational wave antenna to a polarized source".

Tyson, J.A. 1974, IAU Symposium Nr. 64, C. De Witt-Morette (ed.) (Dordrecht: Reidel Publ. Co.), p. 17, "Detection of gravitational radiation."

Vishveshwara, V.C. 1970a, Nature, 227, 936-938, "Scattering of gravitational radiation by a Schwarzschild black hole".

Vishveshwara, V.C. 1970b, Phys. Rev. D, 1, 2870-2879, "Stability of the Schwarzschild metric".

Wald, R.M. 1973, J. Math. Phys., 14, 1453-1461, "On perturbations of the Kerr black holes".

Walker, M., and Held, A. 1974, private communication.

Weber, J. 1960, Phys. Rev., 117, 307-313, "Detection and generation of gravitational waves".

Weber, J. 1961, General Relativity and Gravitational Waves (New York: Wiley-Interscience).

Weber, J. 1967, Phys. Rev. Lett., 18, 498-501, "Gravitational radiation".

Weber, J. 1969, Phys. Rev. Lett., 22, 1320-1324, "Evidence for discovery of gravitational radiation".

Weber, J. 1970a, Phys. Rev. Lett., 24, 276-279, "Gravitational radiation experiments".

Weber, J. 1970b, Phys. Rev. Lett., 25, 180-184, "Anisotropy and polarization in the gravitational radiation experiments".

Weber, J. 1970c, Nuov. Com. Lett., 4, 653-658, "The new gravitational wave detectors".

Weber, J. 1972a, GRG, 3, 59-62, "Advances in gravitational radiation detection".

Weber, J. 1972b, Nature, 240, 28-30, "Computer analyzes of gravitational radiation detector coincidences".

Weber, J., and Trimble, V. 1973, Phys. Lett., 45A, 353-354, "On the response of a gravitational radiation detector to magnetic field fluctuations".

Weber, J. et al. 1973, Phys. Rev. Lett., 31, 779-783, "New gravitational radiation experiments".

Weinberg, S. 1972, Gravitation and Cosmology (London: Wiley & Sons).

Wilkins, D.C. 1972, Phys. Rev. D, 5, 814-822, "Bound geodesics in Kerr metric".

Winterberg, F., and Phillips, W.G. 1973, Phys. Rev. D, 8, 3329-3337, "Gravitational self-lens effect".

Wheeler, J.A. 1962, in Geometrodynamics (New York: Ac. Press).

Wheeler, J.A. 1964, 467-500, in Relativity, Groups & Topology, De Witt & De Witt (eds.) (New York: Gordon & Breach).

Wong, C.-Y. 1971, J. Math. Phys., 12, 70-78, "Application of Regge-calculus to the Schwarzschild and Reissner-Nordstrøm geometries at the moment of time-symmetry".

Zel'dovich, Ya.B. 1971, Pis'ma v. Zh. Exp. i Teor. Fiz., 14, 270 (transl. in Soviet Phys. - JETP (Lett.), 14, 180).

Zel'dovich, Ya.B. 1972, Zh. Exp. i Teor. Fiz, 62, 2076 (transl. in Soviet Phys. - JETP, 35, 1085).

Zel'dovich, Ya.B., and Novikov, 1971, Relativistic Astrophysics vol.I (Stars & Relativity) (Chicago & London: Univ. of Chicago Press).

Zerilli, F. 1970a, Phys. Rev. Lett., 24, 737-738, "Effective potential for even parity Regge-Wheeler gravitational perturbation equations".

Zerilli, F. 1970b, Phys. Rev. D, **2**, 2141-2160, "Gravitational field of a particle falling in the Schwarzschild geometry analyzed in tensor harmonics".

Zerilli, F. 1974, Phys.Rev.D, **9**, 860-868, "Perturbation analysis for gravitational and electromagnetic radiation in a Reissner-Nordstrøm geometry".

Zwicky, F. 1957, Morphological Astronomy, (New York: McGraw-Hill), p. 215.

Lecture Notes in Physics

Bisher erschienen/Already published

Vol. 1: J. C. Erdmann, Wärmeleitung in Kristallen, theoretische Grundlagen und fortgeschrittenene experimentelle Methoden. 1969.

Vol. 2: K. Hepp, Théorie de la renormalisation. 1969.

Vol. 3: A. Martin, Scattering Theory: Unitarity, Analyticity and Crossing. 1969.

Vol. 4: G. Ludwig, Deutung des Begriffs physikalische Theorie und axiomatische Grundlegung der Hilbertraumstruktur der Quantenmechanik durch Hauptsätze des Messens. 1970. Vergriffen.

Vol. 5: M. Schaaf, The Reduction of the Product of Two Irreducible Unitary Representations of the Proper Orthochronous Quantummechanical Poincaré Group. 1970.

Vol. 6: Group Representations in Mathematics and Physics. Edited by V. Bargmann. 1970.

Vol. 7: R. Balescu, J. L. Lebowitz, I. Prigogine, P. Résibois, Z. W. Salsburg, Lectures in Statistical Physics. 1971.

Vol. 8: Proceedings of the Second International Conference on Numerical Methods in Fluid Dynamics. Edited by M. Holt. 1971. Out of print.

Vol. 9: D. W. Robinson, The Thermodynamic Pressure in Quantum Statistical Mechanics. 1971.

Vol. 10: J. M. Stewart, Non-Equilibrium Relativistic Kinetic Theory. 1971.

Vol. 11: O. Steinmann, Perturbation Expansions in Axiomatic Field Theory. 1971.

Vol. 12: Statistical Models and Turbulence. Edited by C. Van Atta and M. Rosenblatt. Reprint of the First Edition 1975.

Vol. 13: M. Ryan, Hamiltonian Cosmology. 1972.

Vol. 14: Methods of Local and Global Differential Geometry in General Relativity. Edited by D. Farnsworth, J. Fink, J. Porter and A. Thompson. 1972.

Vol. 15: M. Fierz, Vorlesungen zur Entwicklungsgeschichte der Mechanik. 1972.

Vol. 16: H.-O. Georgii, Phasenübergang 1. Art bei Gittergasmodellen. 1972.

Vol. 17: Strong Interaction Physics. Edited by W. Rühl and A. Vancura. 1973.

Vol. 18: Proceedings of the Third International Conference on Numerical Methods in Fluid Mechanics, Vol. I. Edited by H. Cabannes and R. Temam. 1973.

Vol. 19: Proceedings of the Third International Conference on Numerical Methods in Fluid Mechanics, Vol. II. Edited by H. Cabannes and R. Temam. 1973.

Vol. 20: Statistical Mechanics and Mathematical Problems. Edited by A. Lenard. 1973.

Vol. 21: Optimization and Stability Problems in Continuum Mechanics. Edited by P. K. C. Wang. 1973.

Vol. 22: Proceedings of the Europhysics Study Conference on Intermediate Processes in Nuclear Reactions. Edited by N. Cindro, P. Kulišić and Th. Mayer-Kuckuk. 1973.

Vol. 23: Nuclear Structure Physics. Proceedings of the Minerva Symposium on Physics. Edited by U. Smilansky, I. Talmi, and H. A. Weidenmüller. 1973.

Vol. 24: R. F. Snipes, Statistical Mechanical Theory of the Electrolytic Transport of Non-electrolytes. 1973.

Vol. 25: Constructive Quantum Field Theory. The 1973 "Ettore Majorana" International School of Mathematical Physics. Edited by G. Velo and A. Wightman. 1973.

Vol. 26: A. Hubert, Theorie der Domänenwände in geordneten Medien. 1974.

Vol. 27: R. Kh. Zeytounian, Notes sur les Ecoulements Rotationnels de Fluides Parfaits. 1974.

Vol. 28: Lectures in Statistical Physics. Edited by W. C. Schieve and J. S. Turner. 1974.

Vol. 29: Foundations of Quantum Mechanics and Ordered Linear Spaces. Advanced Study Institute Held in Marburg 1973. Edited by A. Hartkämper and H. Neumann. 1974.

Vol. 30: Polarization Nuclear Physics. Proceedings of a Meeting held at Ebermannstadt October 1–5, 1973. Edited by D. Fick. 1974.

Vol. 31: Transport Phenomena. Sitges International School of Statistical Mechanics, June 1974. Edited by G. Kirczenow and J. Marro.

Vol. 32: Particles, Quantum Fields and Statistical Mechanics. Proceedings of the 1973 Summer Institute in Theoretical Physics held at the Centro de Investigacion y de Estudios Avanzados del IPN – Mexico City. Edited by M. Alexanian and A. Zepeda. 1975.

Vol. 33: Classical and Quantum Mechanical Aspects of Heavy Ion Collisions. Symposium held at the Max-Planck-Institut für Kernphysik, Heidelberg, Germany, October 2–5, 1974. Edited by H. L. Harney, P. Braun-Munzinger and C. K. Gelbke. 1975.

Vol. 34: One-Dimensional Conductors, GPS Summer School Proceedings, 1974. Edited by H. G. Schuster. 1975.

Vol. 35: Proceedings of the Fourth International Conference on Numerical Methods in Fluid Dynamics. June 24–28, 1974, University of Colorado. Edited by R. D. Richtmyer. 1975.

Vol. 36: R. Gatignol, Théorie Cinétique des Gaz à Répartition Discrète de Vitesses. 1975.

Vol. 37: Trends in Elementary Particle Theory. Proceedings 1974. Edited by H. Rollnik and K. Dietz. 1975.

Vol. 38: Dynamical Systems, Theory and Applications. Proceedings 1974. Edited by J. Moser. 1975.

Vol. 39: International Symposium on Mathematical Problems in Theoretical Physics. Proceedings 1975. Edited by H. Araki. 1975.

Vol. 40: Effective Interactions and Operators in Nuclei. Proceedings 1975. Edited by B. R. Barrett. 1975.

Vol. 41: Progress in Numerical Fluid Dynamics. Proceedings 1974. Edited by H. J. Wirz. 1975.

Vol. 42: H II Regions and Related Topics. Proceedings 1975. Edited by T. L. Wilson and D. Downes. 1975.

Vol. 43: Laser Spectroscopy. Proceedings of the Second International Conference, Megève, June 23–27, 1975. Edited by S. Haroche, J. C. Pebay-Peyroula, T. W. Hänsch, and S. E. Harris. 1975.

Vol. 44: R. A. Breuer, Gravitational Perturbation Theory and Synchrotron Radiation. 1975.

Selected Issues from

Lecture Notes in Mathematics

Vol. 345: Proceedings of a Conference on Operator Theory. Edited by P. A. Fillmore. VI, 228 pages. 1973.

Vol. 346: Fučík et al., Spectral Analysis of Nonlinear Operators. II, 287 pages. 1973.

Vol. 348: A. M. Mathai and R. K. Saxena, Generalized Hypergeometric Functions with Applications in Statistics and Physical Sciences. VII, 314 pages. 1973.

Vol. 358: F. L. Williams, Tensor Products of Principal Series Representations. VI, 132 pages. 1973.

Vol. 361: J. W. Schutz, Foundations of Special Relativity: Kinematic Axioms for Minkowski Space Time. XX, 314 pages. 1973.

Vol. 362: Proceedings of the Conference on Numerical Solution of Ordinary Differential Equations. Edited by D. Bettis. VIII, 490 pages. 1974.

Vol. 363: Conference on the Numerical Solution of Differential Equations. Edited by G. A. Watson. IX, 221 pages. 1974.

Vol. 365: R. P. Gilbert, Constructive Methods for Elliptic. Equations. VII, 397 pages. 1974.

Vol. 373: A. E. R. Woodcock and T. Poston, A Geometrical Study of the Elementary Catastrophes. V, 257 pages. 1974.

Vol. 374: S. Yamamuro, Differential Calculus in Topological Linear Spaces. IV, 179 pages. 1974.

Vol. 376: D. B. Osteyee and I. J. Good, Information, Weight of Evidence, the Singularity between Probability Measures and Signal Detection. XI, 156 pages. 1974.

Vol. 377: A. M. Fink, Allmost Periodic Differential Equations. VIII, 336 pages. 1974.

Vol. 384: Functional Analysis and Applications, Proceedings 1972. Edited by L. Nachbin. V, 270 pages. 1974.

Vol. 388: R. L. Lipsman, Group Representations. X, 166 pages. 1974.

Vol. 394: W. M. Patterson 3rd, Iterative Methods for the Solution of a Linear Operator Equation in Hilbert Space – A Survey. III, 183 pages. 1974.

Vol. 395: Numerische Behandlung nichtlinearer Integrodifferential- und Differentialgleichungen. Tagung 1973. Herausgegeben von R. Ansorge und W. Törnig. VII, 313 Seiten. 1974.

Vol. 399: Functional Analysis and its Applications, Proceedings 1973. Edited by H. G. Garnir, K. R. Unni, and J. H. Williamson. XVII, 569 pages. 1974.

Vol. 408: J. Wermer, Potential Theory. VIII, 146 pages. 1974.

Vol. 409: Fonctions de Plusieurs Variables Complexes, Séminaire François Norguet 1970–1973. XIII, 612 pages. 1974.

Vol. 410: Séminaire Pierre Lelong (Analyse) Année 1972–1973. VI, 181 pages. 1974.

Vol. 413: M. Bruneau, Variation Totale d'une Fonction. XIV, 332 pages. 1974.

Vol. 415: Ordinary and Partial Differential Equations, Proceedings of the Conference held at Dundee, 1974. XVII, 447 pages. 1974.

Vol. 416: M. E. Taylor, Pseudo Differential Operators. IV, 155 pages. 1974.

Vol. 417: H. H. Keller, Differential Calculus in Locally Convex Spaces. XVI, 131 pages. 1974.

Vol. 419: Topics in Analysis – Proceedings 1970. Edited by O. E. Lehto, I. S. Louhivaara, and R. H. Nevanlinna. XIII, 391 pages. 1974.

Vol. 425: P. R. Chernoff and J. E. Marsden, Properties of Infinite Dimensional Hamiltonian Systems. IV, 160 pages. 1974.

Vol. 426: M. L. Silverstein, Symmetric Markov Processes. IX, 287 pages. 1974.

Vol. 427: H. Omori, Infinite Dimensional Lie Transformation Groups. XII, 149 pages. 1974.

Vol. 430: Constructive and Computational Methods for Differential and Integral Equations. Symposium 1974. Edited by D. L. Colton and R. P. Gilbert. VII, 476 pages. 1974.

Vol. 433: W. G. Faris, Self-Adjoint Operators. VII, 115 pages. 1975.

Vol. 434: P. Brenner, V. Thomée, and L. B. Wahlbin, Besov Spaces and Applications to Difference Methods for Initial Value Problems. II, 154 pages. 1975.

Vol. 440: R. K. Getoor, Markov Processes: Ray Processes and Right Processes. V, 118 pages. 1975.

Vol. 442: C. H. Wilcox, Scattering Theory for the d'Alembert Equation in Exterior Domains. III, 184 pages. 1975.

Vol. 446: Partial Differential Equations and Related Topics, Proceedings 1974. Edited by J. A. Goldstein. IV, 389 pages. 1975.

Vol. 448: Spectral Theory and Differential Equations. Proceedings 1974. Edited by W. N. Everitt. XII, 321 pages. 1975.

Vol. 449: Hyperfunctions and Theoretical Physics. Proceedings 1973. Edited by F. Pham. IV, 218 pages. 1975.

Vol. 458: P. Walters, Ergodic Theory – Introductory Lectures. VI, 198 pages. 1975.

Selected Issues from

Lecture Notes in Mathematics